Continental Wrench-Tectonics and Hydrocarbon Habitat

Tectonique Continentale en Cisaillement

Continuing Education Course Note Series #30

By/Par

Greg Zolnai
Elf-Aquitaine
Pau, France

Published March, 1989 by the
Education Department of
The American Association of Petroleum Geologists
Tulsa, Oklahoma 74101-0979

Extra copies of this book, and other titles in the
Education Course Note Series, are available from:

The AAPG Bookstore
P.O. Box 979
Tulsa, OK 74101-0979

ISBN: 0-89181-178-8

AAPG

Mediterranean Basins — Conference & Exhibition
Nice, France

September 25 — 28, 1988

Short Course n° 2

Continental Wrench—Tectonics and Hydrocarbon Habitat

(English Text)

Lecturer: Greg Zolnai

G. Zolnai
Elf—Aquitaine
Pau (64018) France

Present Address
11 Av. des Sayettes
64000 Pau, France

A B S T R A C T

1) In continental interiors most structural elements such as rifts, arches, downwarps, uplifts, collapse-troughs and even folds and fold-groups are due to or involve some form of wrenching.

2) The thickness and rigidity of the crust renders purely compressive deformations with crustal shortening rare. Small-scale compression, extension and shearing can create sizable positive and negative anomalies and extensive fracture-systems. Long strike-slip faults can be generated by moderate shear-motion and maintained by successive reactivations with possible reversals in direction of relative movement.

3) "Paired uplifts and basins" are simultaneously created through the internal reorganization of shear-zones. Without necessarily external compression or extension. Intense tear-faulting and friction along fault-planes induces rotation, tilting, uplifting and second-order compression and extension. Reversal of shear-motion will alter the type and sign of these anomalies.

4) Deformation within the continental interiors will mostly take place along pre-existing suture-belts and zones of weakness; the role of the structural heritage is therefore of prime importance. Mega-shear belts may cross continents without loss of their offset. They can trigger secondary anomalies of various types and sizes all along their path.

5) The first-order fault-systems - geofractures - penetrate the continental crust on its full thickness. They form transcontinental structural webs, which may be rejuvenated by the successive stress-regimes and pick up most of the intracontinental deformation. In the meantime, the rounded "nuclei" they separate will undergo relative adjustments through shearing or rotation.

6) Folds and fold-systems in cratonic interiors (and along passive margins) reflect in most cases draping or dragging along and above deep shear-faults. These are either en-echelon parallel fold-swarms oblique to the master-faults (drag-folds)(, or they are single anomalies parallel to the strike of the main shearing and due to the upthrusting, vaulting or tilting of elongate slabs within the underlying wrench-corridor (drape-folds). None of these necessitates shortening perpendicular to the fold-axes. Such fold-groups can form "passive" orogenic belts of great extent or relief, far away from active orogens.

7) Continents are pulled, rather than pushed, during their drifting; truly compressive belts are limited to narrow strips along their leading edges. Continental interiors will undergo diffuse stretching facilitated by the deep fracture-network and resulting in steadily subsiding depocenters. "Passive tectonism" may in the meantime affect the passive margins with greater degree of freedom.

8) Strike-slip deformation may occur at any period of time during the global tectonic cycle (from rifting and break-up through drifting and collision), although they are frequently synchronous with orogenies. They may reach any part of the continental interiors, since wrench-generated weakness-belts can convey deformation from coast to coast, from the active margins to the passive margins.

9) The direct assessment of the motion along a shear-belt is frequently most difficult. The specific geometry of shear-belts allows the comprehension of the stress-field and the determination of the direction and, sometimes, the amplitude of the displacement, which may be capital in exploration.

10) Shear-zones form preferential habitats for hydrocarbon accumulation since :

   - Great amounts of sediments may cumulate in shear-related collapse-basins and rifts ;
   - The specific paleogeographical settings are favourable for confined basins ;
   - Multiple stratigraphic and structural traps are generated in the areas of tectonic instability during the recurring phases of movements, even in otherwise undisturbed domains ;
   - Extensional fractures create vertical paths for migration allowing hydrocarbons to reach shallow habitats ;
   - Deep fracturing induced by wrenching may improve reservoir-qualities as well as increase heat-flow and allow early maturation.

   The key-factor for commercial hydrocarbon accumulation remains the presence of effective cap-rocks, to prevent dispersion along the ubiquitous open-fractures.

   Due to the different mechanisms involved, the exploratory approach of sheared zones is different of that of compressed belts.

"CONTINENTAL WRENCH - TECTONICS"

- Course outline -

I.	BASICS	1/2 HR		08:00 A.M.
II.	WRENCH - FAULTS VS. BLOCK - FAULTING	1/2 HR		
III.	WRENCH - FAULTS VS. FOLDING	1/2 HR		
			Break	-10:00
IV.	CRITERIA of IDENTIFICATION	1/2 HR		
V.	AMPLITUDE of MOVEMENTS	1/2 HR		
VI	ROLE of WRENCHING in the CONTINENTAL ARCHITECTURE	1 HR		
	discussion	1/2 HR		-12:00 A.M.
	--		Lunch	------------
VII.	HYDROCARBON GENERATION AND HABITAT	1 1/2 HR		13:00
VIII.	WORKSHOP - exercises	1 HR		-15:30
			Break	-16:00
IX.	GLOBAL INTERPRETATION and DRIVING MECHANISM	1/2 HR +		16:00
X.	WORKING RECOMMENDATIONS	1/2 HR -		-17:00
				The End

INDEX

"Everything simplified is false ...
Everything which isn't it is useless."
- *Paul Valery* -

FOREWORD

Continental Interiors are vast, relatively undeformed, stable, and apparently monotonous areas which have long been among the most prolific hydrocarbon producers in the world. They were mostly explored with methods used in tectonically more complex foreland belts, and our exploration philosophy has also been copied from the frankly folded and truly compressed regions along continental margins. With the quickly growing and increasingly accurate exploratory and scientific data, one can attempt to identify the specific characteristics and deformation mechanisms of these interior continental domains and to seek a better understanding of the transition between Orogens and Cratons.

It appears, after the study of a number of cases, that wrenching may have played an important role in shaping the structure of Continental Interiors. Most of our knowledge on strike-slip deformation has, however, been gained in highly disturbed orogenic belts such as ir California and New Zealand. Doubts may thus arise about indiscriminately transferring such mechanisms to the cratonic domains. It is therefore important to identify similarities and differences between the phenomena observed in these different domains, and to quantify their variations with the hope of better understanding the subtle, albeit essential, intracontinental sheared zones.

We may thus not only better understand the fabric of the Continental Interiors, but also consider their largely - as yet - neglected impact on global tectonic processes.

I. BASICS

I-A) INTRODUCTION

The basic diagram of RIEDEL (1927) describes and classifies the fracture-pattern generated by parallel shearing. Of special interest are the synthetic and antithetic fault-groups (Fig. I-5).

HARDING's diagram indicates, furthermore, the disposition of the associated compressive and extensional features, such as fold-axes, styloliths, overthrusts, normal faults and open fractures (joints) (Fig. I-6).

CROWEL's rhomb-diagram summarizes the basic geometry of a faulted pull-apart basin (or lozenge-graben) and the configuration of the most typical secondary features within a shearing system (Fig. I-7).

WILCOX's and KINGMA's diagrams describe the geometry and behaviour of complex shear-zones during the process of on-going shearing (see also § II) (Fig. I-8).

CROWEL and RODGERS have described the scenario for the offset fault-patterns in the process of wrenching (Fig. I-9).

"Transcurrent faults" have long been studied along the western edge of the North American Continent ; the use of the term should be restricted to faults (a) extending as continuous, linear master-faults over several hundreds of kilometers, (b) exhibiting horizontal movements of the hundred-kilometre order (e.g. the Californian type-example, San Andreas fault) (Fig. I-10).

Oceanic "transform-faults" were defined by TUZO WILSON ; their use should be limited to oceanic domains and avoided within the continental areas since the direction of the "throw" of an oceanic transform-fault may contradict the apparent displacement along their wrench-fault extension on the continent and since the mechanisms involved are totally different (Fig. I-11).

The combination of external compression and wrenching results in transpression, which initiates flower structures as well as block-uplifts and faulted ridges. External extension and wrenching creates transtension, generating negative flower structures, as well as furrows, rhomb-graben, faulted downwarps, etc (Fig. I-12/A-B-C).

The effects of wrenching in a given fault-zone or corridor also include rotation of blocks and appearence of open (extensional) fractures. Complex wrench-systems may exhibit gash-faults or en-échelon faults, as well as en-échelon folds. Large-scale shearing frequently results in fault-"lineaments" and may form arches (Fig. I-13).

Deep regional shearing is often expressed at the surface by individual folds or fold-swarms. Drag-folds (or "torsion faults", term proposed by D. RIGASSI) reflect underlying wrench-faulting, whereas drape-folds may hide uplifted, domed or tilted basement blocks or slabs. Both belong to the category of "forced folds" (STEARNS) and do not necessarily indicate crustal shortening (Fig. I-16-17-18).

"Parquet" patterns can be observed in the consolidated, basement-complexes, where elongate, parallel, equally spaced basement-strips undergo wrenching in the same direction (Fig. I-17 & 18).

I-B) FIGURES

Fig. n°

I-C) CONCLUSIONS

- If there is a couple or torque, the initial stress is in the "simple shear" mode ; then (Fig. I-14) :

  (a) the fault-pattern will be asymetrical or non-conjugate,

  (b) the fractures will be most frequently open ("R'" of Riedel),

  (c) the folds will form relays and "en échelon" patterns, their long axis will be at acute angles with the primary force directions /THOMAS/.

  If the initial stress is in compression or extension, not torque but "pure shear", then (Fig. I-14) :

  (a) the fault-pattern will be symetrical and conjugate,

  (b) the fractures will be mostly closed,

  (c) the folds will be in more-or-less symetrical groups or they will form lineaments, whose long axis is normal to the force directions.

- The basic mechanism in hard, competent layers, involves friction along the fault-planes, with subsequent rotation of blocs resulting in "jamming", and consequent local (second-order) compression, hence uplifted blocks and collapsed compartments will appear.

- "Paired uplifts and basins" (J. WALPER) will balance in mass (see below, § II), under translation, while under transpression uplifts predominate and under transtension basins are more widespread. There will be nevertheless always a few blocks of the opposite type.

- In the case of incompetent terrains, the drape and drag-folding is due to inner friction, coherence and plasticity of the strata. Local elongation compensated by "in situ" contraction will be responsible for the appearence of folds. In the process of on-going wrenching, such folds will undergo rotation and may be faulted and overthrusted.

- An uplift or a fold does not necessarily mean compression ; conversely a fracture-system or a downwarp does not mean extension in the regional sense. They may both be due to shearing, which can originate its own local, second-order stress-field with compression and extension, and which will pair positive and negative anomalies. Such structures or structural systems are therefore not due to any shortening at right angles to the "axes", but rather to an internal reorganization of masses, in the process of shearing (see § V. hereafter).

This internal rearrangement of masses will produce different effects in the various structural levels : they will be faulting in the deep, competent domain, while (forced) folding will caraterize the higher uncompetent structural levels. Such tectonic packages may also be superimposed (see chapters II & III hereafter).

WARNINGS :

In a given area compressional and non-compressional folds may be generated simultaneously and at close range (e.g. in orogenic forelands). It is capital for hydrocarbon exploration to identify "which is which ?"

Drape-folds above basement-crests or tilted blocks in a wrench-corridor may be mistaken for drag-folds, resulting in an eronneous estimate of the strike-slip sense (Fig. VIII-7).

One of the problems when applying wrench-tectonics on Continental interiors is that we are usually inclined to think in terms of "pure shear" (i.e. compression) instead of admitting that "simple shear" ("shear" in the common sense) can be the initial force (see below, chapters VI and IX).

II. WRENCHING versus BLOCK-FAULTING

II-A) INTRODUCTION

Wrenching affects competent and incompetent rocks in different ways.

In hard, consolidated rocks, cratonic areas or in the deep structural levels, shearing will induce mainly faulting. In ideal cases, initial en-échelon fracture-systems will merge to form shear-corridors, where rhomb-shaped blocks will emerge between two master-faults. But consolidated domains are usually old platforms where the inherited fabric, the heritage will prevail. The result will be a combination of "expected" (see § I above) and "inherited" patterns with possible interference effects. In other words, the new stress-field may not create its own new structures as much as rejuvenate preexisting ones ; old structures closest in line with new stress-fields will in fact determine the tectonic regime.

A few of the better-known field-examples will be reviewed here, together with interpretations of most operational significance.

II-B) FIGURES

III. WRENCHING versus FOLDING

III-A) INTRODUCTION

In higher structural levels characterized by incompetent strata, the most current structural expression of deep wrenching will be in the form of folds and associated second-order fracture-systems. The folding will be enhanced in case of the presence of highly mobile, soft interstrates, e.g. shales or evaporites near the base of the higher structural level.

Some of the most typical cases of combined wrenching and folding, and superimposed fold-belts and wrench-systems will be considered next.

III-B) FIGURES

IV. IDENTIFICATION CRITERIA for WRENCH-FAULTS

IV-A) In MAPS, wrench-faults can be identified by the presence of one or several of the following criteria :

A) Features disposed at high angles or perpendicular to the major fault-line, such as possibly secondary fault networks, reef trends, channel systems, facies changes, but also intrusive and volcanic bodies and dykes (-swarms) are shifted along, or simply cut off by the wrench-fault (Fig. IV-1-2).

B) Wrenching produces usually narrow, long bands of en échelon features such as parallel but shifted anticlines, gash-faults or both (Fig. IV-3).

C) Anticlines are commonly affected by parallel en-échelon fractures or normal-faults, near normal to the anticlinal axis. The faults do not form (symmetrical) conjugate sets, the fractures are often open.

D) The anticlines are frequently recumbent and sometimes overthrust, on one side or both.

E) Series of folds disposed along both sides of and usually cut off by the master fault, which changes frequently its vergence ("Scissor-faults"). At first sight such arrangements may suggest the presence of a normal-fault trend, with frequently changing sense of throw, rather unusual of normal-fault systems (Fig. IV-4).

F) The presence of large-scale uplifted slabs, more rarely block uplifts, sometimes triangular or rounded "corks" blocking the wrench-corridors may all be taken as indicators of wrench-movements (Fig. IV-6, 9).

G) Single volcanoes are often present along wrench-faults (and -systems) while the associated graben may be filled with volcanic flows.

H) "Paired uplifts and basins" forming elongate checkerboards are typical indicators of large-scale wrench-systems. Normal- and overthrust-faults alternate at close range and the uplifted blocks (or the downfaulted graben) join at angles.

I) In deeply eroded areas shear-zones are composed by groups of (sub)vertical parallel faults, between which a great variety of blocks, amigdaloids as well as strongly squeezed synclines may be present (see below). On the field these belts are strikingly straight, whatever the topography (e.g. the North Pyrenean Fault, shaped through repeated phases of transtension, transpression and compression : BOURROUILH et al. 1988). (Fig. IV-9).

IV-B) In vertical SECTIONS, one frequently observes :

A) Sudden changes of thickness, composition (facies), dip, seismic character of the sequences situated on opposite sides of the wrench-fault (Fig. IV-5).

B) "Flower structures" i.e. "upward-spreading fault-zones whose elements usually have reverse separations" (HARDING and LOWELL, 1983) (Fig. I-12).

C) "Negative flower structures" which "...consist of shallow sags overlying upward-spreading strike-slip faults with normal separations" (ibid.).

D) "Dolphin-effect" (or porpoise-effect, ZALAN, 1987) "between laterally moving blocks" where "the fault plane dips in the same direction in successive apparences (sections) ; however it appears as a normal fault in one section while in the other it is displayed as a reverse fault ; and so on" (Fig. IV-8).

E) "Ribbon effect" : the sub-vertical fault-plane changes its dip and, consequently, its apparent "normal" or "reverse" character (IV-8).

F) "Gaps" on seismic lines, possibly due to the highly complex inner structure of the fault-zone, due to strong cataclasis that takes place, originating uncommon rupture patterns (ZOLNAI, 1975) (ZALAN, 1987) (Fig. IV-7).

G) Allochtonous blocks, erratic rocks commonly from deeper levels (including mantelic material) forming vertical lenses, "amigdaloids" etc., cornered within or squeezed into the fault-plane (Fig. IV-9).

H) Sub-vertical, parallel fault-planes appearing on the same cross-section, some of which are normal-faults others exhibiting reverse separation, are indications of wrench-movements parallel to the fault-plane.

Warning :

- Changes in thickness and character across a wrench-fault line may be due not only to the horizontal shift that took place along the fault at a given time, but also to syn-sedimentary normal-faulting along the same fault during an earlier period ; since fault-lines, once established, may react in different ways and directions during the successive tectonic episode.

- In such cases when magmatic bodies (intrusions, dykes, volcanoes) of similar type and age appear in shifted positions on the opposite sides of a shear-zone or transcurrent fault, one should not jump to the conclusion that the distance between the "half-intrusions" equals the horizontal displacement along the strike of the fault. Such bodies may, in fact be emplaced at the crossing of the master-fault with a (swarm of) secondary fault(s) and such intersections may have been "en baïonettes" or have gained their "Z"-design already before the intrusion took place. In a similar way, reef-trends, channeling sands and even facies-boundaries may "zig-zag", at the moment of their deposition, when crossing a major fault. The strike-slip displacement can therefore be asserted with a certain degree of accuracy only if several types of anomalies exhibit the same sense and amount of movement, which is extremely rare.

IV-C) FIGURES

V. AMPLITUDE of MOVEMENTS

V-A) INTRODUCTION

One of the obstacles to the concept of widespread intracontinental wrench-deformations lies in the difficulty of estimating the amount and amplitude of the actual displacements along a strike-slip fault, or that needed to trigger a mega-shear belt.

- The fundamental problem of the direct observation of the motion related to wrench-faults and shear-systems, lies with the basic architecture of sedimentary basins, our prime targets : we have at our disposal many horizontal markers, (the strata) for each reliable vertical feature... if any. This disproportion may reach several magnitudes (x 10, x 100), since dykes, reef-trends, well-defined channeling deposits and intrusions are scarce and poorly dated. The lack of direct evidence makes that one has to relay frequently on a network of secondary phenomena which may be used as "circumstancial evidence" of the hidden, subtle, elusive, 'though essential phenomenon of wrenching.

- When trying to estimate the amplitude of the strike-slip movements, several obstacles may rise. First, one instinctively associates a large movement to any long structural line (something like one magnitude below the observed length of the fault e.g. 100 km movement for a 1000 km-long fault etc.). Secondly, in the case of "closed" structural features such as pull-apart basins, one expects a movement equivalent to the length of one of the sides of the basin (e.g. at least 50 km of shearing for a 50 x 100 km lozenge-basin). In this case, one wonders how the continents or continental blocks did not fall to pieces when affected by such tear-movements, and one frequently turns to less perillous pure compression-extension mechanisms...

- Fortunately, the results of deep geophysical investigations (seismic, potential methods) did establish the fact that the basic structural lines ("geofractures") affect the continental crust in all of its thickness (e.g. upwelling of the "Moho" discontinuity below the rifts). This means that the mass of the whole lithosphere has to be taken into account when interpreting the major anomalies, instead of considering them as mere skin-effects of the crust. We can thus find a key to the quantitative estimation of the horizontal movements, provided the volume or geometry of the anomalies can be established.

  In this chapter, a few simple geometric schemes are presented and applied on observed (field-) anomalies.

V-B) FIGURES

V-C) CONCLUSIONS

The "modesty" of the horizontal displacements required to create important anomalies is why the role of wrenching in the shaping of the continental interiors has been neglected or underestimated for so long. Such modest shifts are indeed very difficult to spot, to measure or to estimate.

- Small to moderate horizontal movements (1-10 km magnitude) in the extension or compression may generate sizeable basins or spectacular uplifts, rifts or arches, intracontinental aulacogens and "miogeoklines" or mountain-ranges, provided the crust is thick which is mostly the case in continental interiors (30 km min., 50 km max.) (Fig. V-1-5).

- If shearing occurs, the horizontal and vertical throws will be similar but the anomalies they generate may be more abrupt (see below).

- The thickness of the underlying crust is a key factor in the quantification of such movements (Fig. V-6) :

  . if the crust is thick, small (1 km order or even less) horizontal movement, in compression or extension suffices to create vertical anomalies 5 or 10 times greater (e.g. in the "bends" or "offsets" of wrench-systems) ;

  . in the case of thin continental skin, a given wrench-movement creates pull-apart basins of similar length (or width). In this case, the depth of the basin will be close to that of the crust and oceanic material will be found at its bottom (Fig. V-6).

- The rigidity of the cratons (their lateral un-compressibity) makes that the width of the wrench-systems (corridors) remains constant during the strike-slip deformation. The blocks, taken into the process of wrenching will therefore (owing to the drag-effect of the neighbouring platforms) undergo rotations, "jam" and eventually surge, leaving collapse-graben between them and along the border-faults of the shear-corridor. It is nevertheless impossible to establish a geometrical rule between the strike-slip and uplift component since in simple geometry there is no friction, thus no rotation/jamming and no uplift can be visualized (Fig. V-7-8).

  Based on a few, rather controversial cases (like the Uinta uplift, in Utah-USA), one may suggest that the horizontal and vertical movements may be of the same magnitude, the horizontal component being possibly 2-3 times greater than the vertical one (Fig. V-10).

- Although pure compression and extension may occur in continental interiors, the most frequent deformation mechanism remains wrenching. In these cases, the "positive" and "negative" anomalies are due to an internal reorganization or rearrangement of the masses taken into the process of shearing, rather than to crustal shortening or elongation (Fig. V-9).

- If the basement is rigid and thick (i.e. uncompressible) the same amount of wrenching displacement may be projected over long distances without loss of its "arrow". Attenuation (tapering off) may nevertheless occur when (and where) compressional <u>and</u> extensional structures branch off, on the opposite sides of, the main tear-fault (Fig. V-11-12).

- The <u>drag-folds associated to wrenching</u> are not due to regional compression either, and the fold-axes are not perpendicular to the principal (shearing-) stress but are oblique to it. The re-arrangement of the masses takes place in this case within the sedimentary layer or sequence, through directional thinning and sagging. This "transfer of masses" is very hard to observe in the field and virtually escapes sub-surface observations such as drilling or seismic, since one cannot follow the detail of the variations of the thickness of the layers all along a given structure).

  Sandbox experiences by F. ODONNE have demonstrated the early appearence of drag-folds during the process of wrenching of the "basement" and also the fact that those folds change their orientation while the wrenching increases, to become more-and-more parallel to the master-fault. The fold-axes are increasingly recumbent away from the middle-line and the set of the fold-axes diverge upwards ("bouquet de fleurs") (Fig. III-6).

- The <u>mass-balance</u> is usually <u>positive</u>, since :

  a) shearing usually introduces some porosity (fracture or possibly matrix), leading to the "swelling" of the wrench-corridors (experiments of BARTLETT et al.) - Fig. V.13.

  b) when pull-aparts are filled with detritus derived from nearby block-uplifts, the sedimentary infill is much looser (a = 1,5 $\pm$) than the hard rocks being eroded (d = 2,7 - 3) ; hence the "left-over" material bridges the gaps between the isolated uplifts thus creating a continuous ridge ;

  c) the positive mass-balance may be further enhanced by external factors. Deep fracturing can facilitate intrusions (granites) and initiate crustal thickening by cristallization, both increasing the buoyancy of the wrench-segments and by consequence amplify the uplifting by isostasy. The erosion is thus compensated by steady "epeirogenic" uplifting. In addition, the vertical, mostly open tear-fractures will help volcanism to take place, adding both "mass" and "relief".

- The basic rule, when considering intracontinental (or trans-continental) wrench-systems is nevertheless :

  "Rid yourself of the San Andreas Syndrom".

VI. ROLE of WRENCHING in the CONTINENTAL ARCHITECTURE

VI-A) INTRODUCTION

In Continental Interiors, many observed structures are related to some sort of strike-slip deformation. These are also often hydrocarbon-bearing.

Wrench-movements may affect any part of the inherited structural fabric but most frequently will be concentrated into narrow, complex zones of weakness. These zones are connected and thus form a primary network of basement anomalies and fracture-belts, which zig-zag or meander between the individual cratonic blocks, "nuclei" or mini-plates. They are spaced more or less regularly around blocks of a diameter in the order of 500-600 km.

These zones of weakness are composed of connected arches, lineaments, geofractures, linear uplifts, rifts and aulacogens, and may also include hingelines, separate pull-apart basins and isolated uplifts. They can carry deformation across the continents, from the active margins to the passive margins and are able to trigger anomalies along their path.

All these zones are genetically related and display a certain amount of strike-slip movement, through the repeated phases of tectonism, and can respond and re-adjust to successive stress-fields. Thus (Fig. VI-1) :

. pull-apart basins may be changed into uplifts ;

. strike-slip faults may become dip-slip or even reverse ;

. earlier normal-faults along the edges of collapse-(rift-)systems will become preferential sites for later strike-slip displacements ;

. arches may be rifted or collapse altogether, as well as

. huge blocks may be extruded from within collapse-systems (aulacogens) ;

. hinge-lines may become (or merge into) rift-systems, or, conversely, form structural barriers especially near the edges of continents ;

. intracratonic checkerboard and radial systems can be reactivated, into tilted blocks, parquets or keyboard ensembles, through simple extension and/or wrenching.

Most -if not all- large intracratonic downwarps, depressions and depo-centers seem to coincide with and relate to underlying rift (aulacogen) systems (e.g. Williston, Hudson's Bay, Illinois, Paris/France Basin etc.). Their periodical or steady subsidence is accounted for by preferential stretching of the crust at or near the crossroads of zones of weakness or sutures. One may call these downwarps : fontanels.

Wrenching and normal faulting may be present at deep structural levels, during phases of subsidence at shallower structural levels, and fracturing with above-average heat flows may ensue. Sharper "funnels" appear in orogenic forelands where foldbelts intercept fault swarms and rifts of the underlying shield (Fig. VI-2).

Some of the previously homogeneous continental "nuclei" may be split during later tectonic activity (e.g. transtension and rotation) and give place to new intracratonic graben and rifts, which usually widen towards the border of the craton(*).

Passive margins may display features similar to those of the interiors of the continents ; deformation is usually greater, and may be due to the greater mobility of the continental margin crust, thinner and laterally more mobile. Typical faults along stretched trailing edges are normal, listric faults, while sheared, passive margins typically display transtensive and transpressive wrench-faults (e.g. the Arctic Foldbelt, Canada : Fig. VI-10).

Widespread but little studied and poorly understood are the peripheral graben or marginal rift systems, formed and subsequently deformed under a combination of extensional and shearing stress, along and parallel to the passive margins. These elongate, faulted basins are filled with usually thick sedimentary sequences of almost any type, depending on the source areas, and their confinement. Besides clastics and carbonates, they frequently contain evaporites, proving that these troughs or furrows were closed from open sea. Such peripheral graben may also become preferential habitats of wrench-deformation, if submitted to a different stress-regime. They may also be incorporated into orogenic belts, after a complete reversal of the global tectonics (Fig. VI-4).

Continental blocks may be calved from the passive margins, and induce considerable tectonic disturbance in the form of wrenching or transpression related to the rotation of the detached landmasses. Very complex shearing patterns may thus be generated, accompanied by secondary compression, uplifts, drape or drag-folds, enhanced by halokinesis, and basic volcanism. (The absence of penetrative metamorphism, large-scale acidic intrusions, arc-type volcanism and ophiolitic masses lead to believe that this type of "passive tectonism" stands apart from the "true" orogenic processes, of crustal shortening, subduction, and/or collision along the active margins) (Fig. VI-10-11).

Wrench-faults and wrench-systems are numerous and varied in active-margin areas and orogenic belts. The role and behaviour of these are beyond the scope of this short course ; but a quick overview has to be included here, in order to (a) tie in these wrench-systems to those of the continental interiors and (b) to show how their later development may affect the stable platforms and eventually the shields (Fig. VI-6-9).

(*) e.g. the Paradox Basin in the Colorado Plateau and the Parentis Basin in the Aquitain Block, S.W. France.

- In <u>Ocean-Continent</u> collision zones, "long-shore" transcurrent faults are extremely frequent and well documented (California - Alaska, Indonesia, New Zealand), and their displacement is several hundreds of kilometers. Similar but more modest in terms of displacement, wrenching also occurs parallel to the axis of the fold-systems (e.g. the Tintina fault in the Canadian Rockies, the Kaltag fault in the Alaskan Cordillera etc.). Transverse wrench-faults may cross the whole width of Orogens : Lewis-and-Clark and Texas lineaments in the case of the North American Cordillera. The "arrow" of movement along these shear-zones is even smaller, in the 1-10 km order.

- In <u>Continent-Continent</u> collision zones longshore faults may be absent or masked by later tectonic events. The "intra-montane" wrench-faults are nevertheless very well developped, slightly post-orogenic, and opening successor basins (pull-apart graben). In the North Appalachians (Caledonides) this faulting is Carboniferous, in the Southern Appalachians (Variscan equivalent) it is Triassic.

- In the <u>Intra-Continent Convergence Belts</u> (Intracratonic Orogens) wrench-systems exist both parallel and perpendicular to the mountain-axis. The Pyrenees are a typical multi-phase intracratonic Orogen, where the North-Pyrenean fault runs parallel to the axis of both Hercynian and Alpine structures ("N-110°") ; it has been active during the Mesozoic (Cretaceous) phase of "geosynclinal" evolution (or extensional phases) and constituted the southern rim of the basin-development. Late -and post-tectonic N-S trending "transverse" faults are also known to be present in the Pyrenees and their foreland (most of which exhibit shear-movement : ZOLNAI - 71, 75).

- Some of the best-known foldbelts correspond in fact to <u>orogenic forelands</u>, and are underlain by cratonic (craton-edge) platforms : e.g. the Folded Jura Mountains in France and Switzerland, the Franklin Mountains in N.W. Canada and possibly the Saharian Atlas foldbelt in Algeria. The basement has been, in these cases, re-mobilized through wrench-movements along parallel "parquet"-faults. The folding is not (principally) due to the effects of the compression/obduction of the main Orogens, but to small-scale (km-order ?), adjustments between the basement blocks of the cratonic foreland, expressed by drag-folds in the mobile sedimentary cover.

- Another category of orogenic or mountain-belts are related to <u>aula-cogens</u> (mega-rifts), oriented perpendicularly to and opening towards an active margin. These have been <u>re-mobilized</u> by <u>wrenching</u> during or after an orogenic compression or collision against the active margin. Typical examples are the Ougarta "Mountains" in the Northern Sahara, a hercynian structure, and the "East Arm System" in the North West Territories of Canada. Shearing along the faults of the aulacogen create drape and drag-folds, and second-order (gravitational) nappes occasionally. The amplitude of these strike-slip movements is in the order of 1-10 km (Fig. VI-14).

- In the figures VI-12 to 22 a few typical cases of composite shear-zones are presented.

VI-B) FIGURES			Fig. n°
1) Continental Interiors			
. Lineaments Arches Rifts Aulacogens Uplifts	(Evolutionary scheme of inter-related tectonic elements with wrench-components)	/G.Z./	VI-1
. Intra-cratonic depression - "fontanels" Foreland collapse-holes - "funnels"		/G.Z./	VI-2
. Shear-Corridors, Uplifts and Basins		/RUHLAND/ /G.Z./	VI-3
. Hinge-lines & Peripheric Rifts		/G.Z./	VI-4
. "Chessboards"		/THOMAS/	VI-5
2) From active margin to passive margin			
. Evolution of the North American Continent		/G.Z./	
	Mid-Proterozoic		VI-6
	Late-Proterozoic		VI-7
	Paleozoic		VI-8
	Mesozoic-Tertiary		VI-9
. A passive-margin orogeny : Artic Islands/Canada		/G.Z./	VI-10
Boothia Uplift : Artic Islands/Canada		/OKLULICH/	VI-11
3) Examples of composite wrench-systems			
. Liard - Rainbow - Athapuscow - Amer belt Peace River - Athabasca Lake (N.W. Canada)		/G.Z./	VI-12
Liard Basin		/G.Z./	VI-13
Athapuscow aulacogen		/HOFFMAN/	VI-14
Amer Belt		/PATTERSON/	VI-15
Peace River Arch/Trough		/G.Z./	VI-16
. Belt - Big Snowy - Black Hills uplifts Snake River - Billings - Weldon faults (W. USA)		/G.Z./	VI-17
Sumatra - Cat Creek section		/G.Z./	VI-18

. The "Ancestral Rockies" system,	/G.Z./	VI-19
late Paleozoic (W. USA)	/U.S.G.S./	VI-19/B
"Ancestral" vs. "Laramide" Rockies	/G.Z./	VI-19/C-D-E
. Britanny (W. France)		VI-20
. The Central Australian Trough (Fitzroy - Amadeus)	/G.Z./	VI-21
. Western Indonesia - Malaysia	/G.Z./	VI-22
. Eastern France (Burgundy, Jura, Rhine Rhône Areas)	/RUHLAND, GELARD, BERGERAT,G.Z./	VI-23

4) Consequences - "Side-effects"

. Diapirism - halokinesis (Franklinian furrow, Arctic Isl.)	/FOX/	VI-24
. Gravity image (U.S.A.)	/G.Z./	VI-25

VI-C) CONCLUSIONS

- One of the most significant results or recent deep seismic investigations (COCORP, LITHOPROBE, ECORS etc.) is the depiction of rifts, aulacogens and related pull-apart features corresponding in depth to upheavals of the "Moho" discontinuity which can be interpreted as new evidence of the stretching of the lithosphere along those downwarps. This has already been suggested earlier by the characteristically positive (heavy) gravity and magnetic anomalies along the rifts etc (Fig. VI-25).

- Many of these faulted (graben) systems appear, at the time of their formation (i.e. during the incipient rifting phase), to run normal to orogenic belts which they usually hit at specific points, where the continental edge is bent. In the case of most recent rift systems (East Africa-Red Sea-Dead Sea, Western Europe) there is evidence that strike-slip movements occured during the rifting and that the latter were predominant at least in some of the systems' segments. The appearence of such rift-systems, recent or old, is always related to in space, and coincident in time with major Orogens (collisions).

  Once they are emplaced these rift/wrench systems remain mobile and react promptly to the later variations of the regional stress-fields, through extension, wrenching and compressional deformations. Rifts thus become aulacogens, which may be partially or completely inversed into uplifts, etc. ; all these deformations involving various amounts of shearing.

- Arches (s. as the Transcontinental, Tathlina and Peace River arches in North America) remained positive during very long periods (several hundreds of m.y.). This can be easiest explained by them having deep "roots" (or downwarps of the base of the lithosphere), gained most probably through compressive, transpressional deformation. In fact several such arches fall into line with known shear-zones (Tathlinia arch vs. East Arm system in n.w. Canada). They exhibit diffuse negative gravity anomalies.

- Some of these arches may be rifted during later extensional/transtensional deformations ; but all rifts did not exhibit arching or (thermal ?) doming immediately preceeding their rifting. Many of them have been up-tilted at their rims only after their emplacement, possibly in relation to a later translational (wrenching) process.

- The increasing amount of available field, geophysical and subsurface data allow us to connect the major lines of anomalies into transcontinental systems. Most of them show direct evidence of shearing and all are better -or more simply- explained with some shearing component. Such mega-shear zones are essential parts of the continents' structural fabric which carry deformation from the active margins into the inert continental interiors or right across to the passive margins.

Adjustments along such transcontinental systems may allow us to better explain the smaller-scale intracratonic anomalies, most frequently situated along these belts of weakness or branching off them (s. as isolated uplifts, block-faulted areas, pull-apart basins, but also depocenters and isolated folds or fold-groups).

Such structural networks mean that changes of the external stress-field may generate non-compressional folds almost anywhere in the continental interior.

- Each new stress-field (phase of tectonic deformation) may add new elements to the basic continental fabric, usually in form of lineaments or rifts-systems. Some of the inherited structures may in the meantime remain dormant during one or several successive diastrophisms.

- The various types of anomalies, composing at a given moment the connected segments of a mega-shear may also be the successive stages in time of the evolution of each of the segments.

- Little horizontal "shifts" (wrenching, transtension, transpression) may initiate sizeable vertical anomalies, repeated over long distances, since there is no necessary attenuation ("loss of arrow") upon each anomaly of the shear-belts. Conversely, a long wrench-fault (shear-system) does not necessarily indicate a great amount of horizontal displacement, and what displacement happened along it may not have been in the same direction all of the time.

- Young faults often hide old basement-features at depth. Superficial or shallow offsets in fold-trends may thus be due not to recent, post-fold wrenching, but they may express old, pre-existing basement-fault arrangements (the drape-folds are expressions of rejuvenated but previously-wrenched basement-fault systems : e.g. Laramide folds in the Wyoming Province).

- Transcurrent faults usually situated along an "active margin" and having undergone important (N x 100 km) strike-slip motion may, after collision, become intracontinental fault-systems or sutures. These may remain weakness-belts or mega-shears with a renewed but less ample wrench-component (N x 10 km). Even later the same lineament may undergo modest strike-slip deformation (of the km-order of smaller) or yield to extensional or compressive stresses and become graben, pull-apart basin or trough/aulacogen. With age, some of the wrench-systems decrease progessively their mobility, unless a new and very powerful deformation starts a new "cycle".

- The most complex, and less studied aspect of the transcontinental megashears is their impact on the structure of orogenic belts or active margins. It has long since been observed that some of the fundamental fault-complexes or lineaments (Texas, Lewis & Clark), radiating out of the continent's interior -or crossing the whole craton- do not stop at the leading edges of the overthrust-belts but they can be followed, in a subdued manner, all across the orogenic complexes :

  . from craton to craton in the case of intracontinental foldbelts (Pyrenees) ;

  . right through the foldbelt and into the oceanic domain in the case of pericontinental orogens (North American Cordillera).

  At the intersection of these fundamental faults, the structural style, often changes (polarity, frequence of folds, etc.) ; the leading edge of the overthrust belts is bent or shifted and oceanic ridges are developped in their seaward continuation. The emplacement of some of the "terrane" complexes seems, likewise, to be related to the position and direction of those transcontinental shear-zones, making them even more "suspect"...

VII. HYDROCARBON GENERATION AND HABITAT IN RELATION TO WRENCH-FAULTS

VII-A) PROS and CONS

PROS

A) Syn-sedimentary ("slow") wrenching usually contributes to complex, unique paleogeographical and paleotectonic settings, such as :

- . more-or-less closed pull-apart basins,
- . embayments, sea-arms along rejuvenated old features (rifts, aulacogens, fault-swarms),
- . keyboard structures, running out to passive margins,
- . checkerboard patterns with tilted basement-blocks.

Most of these will result in some degree of confinement giving place to the generation (abundant algal life), accumulation (euxinic environments, cul-de-sac or "funnel"-type downwarps) and preservation (no transcurrent sea-streams) of organic material. Slow subsidence rates and lack of clastic influx, during the early stages of the basin-forming cycle may result in extremely rich, concentrated source rocks. Subsequent fine clastic (shaly) to evaporitic sedimentation may provide excellent cap-rocks as well as contribute to the dolomitization of carbonates if there is any. Structural complexity of the graben-rims ("staircase" arrangements) may help reefal build-ups. Ideal conditions may thus be created for hydrocarbon accumulations, especially if the fracture-networks open migration-paths at the right moment (Fig. VII-1-2).

(It should be reminded here that such closed systems may host mineral accumulations as well ; e.g. Lady Loretta mine in Central Australia).

B) Syn-tectonic (intense) shearing, combined with slight extension may result in extremely high subsidence-rates, with the result of great thicknesses of sediments being cumulated, the pull-apart etc. basins acting as sediment-traps vs. nearby - or remote - orogenic belts. In the case of some of the fault-controled basins the estimation of the total sedimentary colum is still uncertain. (e.g. Fitzroy trough in W. Australia, Anadarko Basin in the S.W. USA).

C) Since shearing occurs most frequently along re-juvenated old structural lines, incipient habitats, "synchronous highs" (SCHOLTEN) may exist, at the right moment to trap hydrocarbons, while they are being generated.

D) Syn-tectonic deformation may be projected, along wrench-systems, mega-shears, etc., into the "inert" craton-areas, far away from the better-known disturbed belts. Structural habitats in the form of elongate arches as well as large flat domes may thus be generated "in the middle of no-where" (Fig. VII-3 to 5).

E) Since wrench-faults have often repeated life-periods, various types of habitats may be generated in the same vertical column. Stacked habitats may also be due to the fact that shear-faults (being mostly vertical) cut across the whole sedimentary column, initiating closures at many different depths (Fig. VII-6).

F) Slight wrenching may trigger both volcanism (laccoliths) and halokinesis, both generating structural traps.

G) The ubiquitous extensional fractures (present even under trans-pressive circumstances) may :

- open migration-paths, connecting the usually deeper source-rocks with the overlying reservoirs, contributing to form shallow accumulations with mature oils (Fig. VII-6-7) ;

- increased heat-flow may enhance maturation, even create early and shallow maturation ;

- the process of wrenching promotes fracturing, thus increasing the volume of porosity, and may generate sizeable reservoir-bodies in basin-bottom (synclinal) positions, where old rifts are frequently hidden (e.g. the Northern Uinta and Central Williston Basins, U.S.A.).

CONS

A) The extensional, or open fractures represent an extremely real dangerfor the hydrocarbons to escape ; either at the time of their formation (dismigration to the surface) or at a later stage (Fig. VII-8).

B) The old weakness-belts, due to their repeated tectonic reactivations, are very vulnerable to erosion.

C) The increased heat-flow, while capable to generate conditions of maturity at shallow depth, may cause over-maturation at otherwise normal depth.

D) The thick sedimentary sequences, dumped in to quickly-subsiding basins and supplied by nearby active uplifts, may be chaotic ; the clastic bodies immature and discontinuous, sands ill-sorted. Strong syn-sedimentary tectonism may also cause structural closures to be widespread but small in size. Dispersed habitats are frequent in such quickly-subsiding basins.

VII-B) <u>FIGURES</u> Fig. n°

VII-C) CONCLUSIONS

The key-factor of hydrocarbon-accumulations in wrench-controlled areas is good cap-rock. Explorers must pay attention to whether escapement-routes cannot be detected (on seismic sections etc.). On the other hand any reservoir, even the very shallow ones, may be filled with light, mature oil, generated in depth (in such cases gas is usually absent).

"Old" wrenching is usually good : it may have helped habitats etc. to be formed ; "new" wrenching may be dangerous, because of the possible dismigration.

Limited amount of structural closure at a level may be compensated by the stacking of habitats.

The search of "step out" pools may be different in areas where wrenching predominates from areas with truly compressive tectonics ("shifted" or "en échelon" targets, instead of "lineaments" or "parallel folds").

The mechanism and timing of the process of maturation - migration may be difficult to seek and very different from target-areas nearby.

VIII. WORKSHOP - EXERCICES

1) "CREASE of the TROUSER"
(How to check the orientation of drag-folds vs. the direction of wrenching, safely and - without being noticed !...).

2) FAULTS and FOLDS - experiment :

- synthetic wrenching VIII-2A
- antithetic wrenching VIII-2B

. Take a cutter and make slots along the oblique, en-echelon lines, between the two master-faults of the wrench-corridor ;

. Observe the displacement of the half-dots when "wrenching" the sheet as indicated on the edges of the wrench-corridor ;

. Try to seek the appearence and displacement of the folds which form in the meantime.

3) "WHAT AN INNOCENT FOLD CAN HIDE"

Draw different deep deformation-mechanisms, able to create identical or similar superficial anticlinal forms.

Solutions : VIII-3/B-C.

4) "Where is the MASS-BALANCE zone ?"

. Why do we only meet "pop-up" structures in shear-zone "a" ?
. Why do we only meet "pull-apart" features in shear-zone "b" ?

Solutions : VIII-4/B-C.

5) THINNING of the crust through STRETCHING

. How much lateral movement (wrenching) is needed to create a pull-apart basin of 20 km x 20 km x 10 km, supposing the crust's thickness is : 20 km ? or 30 km ? or 40 km ?

. Draw more realistic basin-profiles in the space at the bottom of the page.

Working sheet : VIII-5/B
Solutions : VIII-5/C

6) Flattening out of folds or "HOW MUCH IS TOO MUCH" ?

. How much crustal shortening was needed to create the :

- drape-fold "a" ?
- drag-fold group "b" ?

Solutions : VIII-6/B

7) DRAPE FOLD OR DRAG FOLD ?

Taking as example the case "a", draw the direction of wrenching for the case "b". Are there any alternatives ?

Solutions : VIII-7/B.

8) LOZENGE - BASIN

. Which stress-field(s) can create the given pull-apart basin ?

. Which stress-field can initiate a compressive fault-pattern with similar directions ?

Solutions : VIII-8/B.

9) SCISSOR - and - SCOTCH-TAPE exercice

. Take the figures VI-19/C-D-E (or make copies of them) and,

. Piece them together to form a "TRIPTYCH" ; then,

. Study the behaviour of the various fault-groups during the successive tectonic phases ("Ancestral" and "Laramide" Rockies).

This area is the clearest demonstration of why "structural geology is a four-dimensional discipline" /D. STONE, 1974/

(You may compare the above triptych with figures VI-6 to 9, representing the continental framework).

VIII. WORKSHOP - EXERCICES

1) "CREASE of the TROUSER"

2) FAULTS and FOLDS - experiment :

3) "WHAT AN INNOCENT FOLD CAN HIDE"

4) "Where is the MASS-BALANCE zone ?"

5) THINNING of the crust through STRETCHING

6) Flattening out of folds or "HCW MUCH IS TOO MUCH" ?

7) DRAPE FOLD OR DRAG FOLD ?

8) LOZENGE - BASIN

9) SCISSOR - and - SCOTCH-TAPE exercice

IX. GLOBAL INTERPRETATION AND DRIVING MECHANISMS

- Continents are not monolithic blocks but rather mosaics, divided by belts of weakness, which are an important part of their inheritance. The latter are the loci of significant stress and deformation. Superficial stray structures dispersed across cratonic domains are most often related and may indeed be good clues to the whereabouts of such deep-seated zones of deformation (Fig. IX-1).

  Although some of the segments of the above structural pattern may respond readily at any moment to any new stress, new anomalies ("shortcuts") are added to the old network, mostly in the form of lineaments, geofractures or rift-graben. These are generated under shearing stress, within the blocks ("nuclei") or mini-plates composing the continental mosaic which are not immutable either. They have their internal structural patterns that may favor further break-up under added stress.

- Deformations within the continental interiors are not triggered by coast-to-coast orogenic compressions ; they are instead initiated by slight adjustment-movements in the wrenching mode between more-or-less loosely bound continental blocks. Orogenic compression or crustal shortening remains limited in most cases to a narrow strip along the active margin, directly exposed to the collision. In contrast, blocks along the sheared margins or the trailing edge of the continent may undergo rotation and eventually break away. All these intra-plate adjustments imply at least some wrenching (Fig. IX-2-3).

- Continental drift is driven at astenospheric levels by convection-cells associated with individual blocks or mosaics ; the driving force must be "self-contained" for each continent. It acts in a "front-end-drive" way, as if the leading edges of the convestion-cells were lined up, to match along the active continental margin with the result being (Fig. IX-2-4) :

  a) the bulk of the blocks or mosaic-pieces away from the leading edge are pulled by forces at the leading edge, and from underneath (rather than pushed across the crustal section, a current misunderstanding) ;

  b) the majority of continental blocks are most of the time under extension. Depocenters will be predetermined by preferential crustal stretching, and therefore sagging, along shear zones and their intersections ;

  (The continuity of subsidence in such intracontinental basins may have been one of the factors hindering the acceptance of the concept of continental drift. The downwarps were associated to deep, "vertical" processes which would have been difficult to shift together with the wandering continents).

c) the immersion of the various continental blocks being different, their response to the continental drifting momentum (or drag-effect of the under-currents) will be different. Shearing and centrifugal motions -with rotation- may be generated amongst blocks (Fig. IX-5) ;

d) wrenching and rifting may occur as soon as the drifting motion of the continent has started. The major deformation will be diachronous from active-margin collisions, continentward to shearing between mosaic blocks ;

e) intrablock adjustments may, however, reverse the sense of shear with respect to the initial impact or incipient rifting event. This may lead to contradictory stress patterns within the globally diachronous setting, such as simultaneous extension and compression among adjacent blocks (Fig. IX-5).

. A global crustal motion will create one leading edge and one trailing edge, with various amounts of jostling inbetween ; it is capital to reveal separate tectonic regimes in their synchroneity or diachroneity. The mass and inertia of the continents is such that they will not change suddenly the direction of their motion ; no "flip-flop subduction" should therefore be envisionned (BLAKE C.M.Jr., 1985).

. Collisions do not always take place along straight fronts moving in exactly opposite directions ; they rather occur along salients and rentrents and may incorporate foreign blocks ("terrane") as well as some broken off their original composition ("lost soldiers"). The potential for shearing among autochtonous, parautochtonous and allochtonous blocks in therefore great. Superimposed upon intra-block readjustments this opens the way to trans-orogen shearing and possibly suturing, in patterns which are totally non-linear in space (i.e. crooked) and time (i.e. tectonic reversals). See the theories on this matter of PAVONI (1961) and TAPPONIER & MOLNAR (1976).

. The weakness-belts will react in different fashion to be successive external stresses or "sollicitations", by changing the direction of their horizontal and vertical movements, the intensity of which may decrease. Some of them may even remain dormant for a while ; but

"Old Faults Never Die" (W.W. MALLORY, 1982).

. The mechanisms themselves, at the origin of observable structures, far better explain intracontinental tectonic regimes than thermal effects resulting from those mechanisms. Crustal thinning will induce heat-flow increases, wrenching may facilitate intrusions and extrusions and intersecting shear-zones may witness crustal cooling, and not the other way around. Especially in the case of recycling events, it is capital to keep structures, mechanisms and timing in clear focus, and to see how they may relate to each other across continental domains.

. During their drifting the Continents may loose <u>break-away</u> blocks along their sheared margin, as well as <u>agglutinate</u> foreign continental masses (terrane) along their active margin. Such detached landmasses may retain some of their structural fabric and they may influence the tectonic evolution of the orogenic belt into which they were incorporated, by forming intramountain swells, as well as triggering early folding along their edges. The old shear-systems may thus re-appear, in a subdued but real way, within the recent foldbelts.

IX-B) FIGURES

X. WORKING RECOMMANDATIONS

- Always work in THREE DIMENSIONS -make block-diagrams rather than maps or sections alone.

- Take into account the "FOURTH DIMENSION" : TIME. Try to unveil the timing of the successive events, do not jam all deformations into one only (recent) period of deformations. Always distinguish what is old (HERITAGE) and what recent movements have added or modified.

- Think REGIONAL. Most hard-to-understand "local" problems can be best solved if replacing them into the regional context : an area one magnitude greater. Don't hesitate comparing with next-door basins, rifts, structures etc. similar features, even if those happen to be geographically remote.

- Use ALL METHODS possible, recent or old, expensive or "cheap". Even such as potential methods (gravity, magnetism), refraction seismograph, morphology and -last but not least- field observations.

- READ all the relevant papers, look at all the maps, including the purely scientific ones or those generated by the mineral industry. No information is useless or dead information.

- Keep an eye on BIG BROTHER BASEMENT ! Try to follow it on all your seismic profiles, even if it necessitates re-interpreting or re-processing.

- Modelling is good, sometimes necessary and unavoidable. But ANALYSIS and pragmatical THINKING is better or -at least- complementary. Don't fall into the trap of over-sophisticated treatments (although they may provide you with great intellectual satisfactions) before you have exhausted all the available down-to-earth information (although this may be boring).

- Do take into account all the available information, but DO THINK INDEPENDENTLY "If you want to reach the source, you have to move against-the-stream" (VOLTAIRE).

XI. ACKNOWLEDGEMENTS

I thank the "DIRECTION EXPLORATION" of the S.N.E.A.(P.) for their material aid in the preparation of this Short Course.

I also wish to express my gratitude to many of my former collegues and long-time friends with the ELF-Aquitaine Group and subsidiaries who, ghrough their advice and discussion -or simply by listening- helped me to piece together my philosophy and realize this report. I especially wish to thank J.P. RICHERT and members of the Structural Geology Group (PAU). J. GIRAUDMAILLER, P. VERNHES, J. HENRY and R. RICATEAU helped me through my early endeavours in the subject : Sylvette TRIBOULET, J.R. STEVAUX, J. de la SERVE and G. NELY also gave their assistance in improving the successive French versions.

During my trips to North America and the time I spent in Canada, J. HAUN, W.W. MALLORY, R. de VOTO, Robbie GRIESS and Annie COURY (all from Denver) as well as P. HOFFMAN (Ottawa) assisted me throughout various discussions and provided me with precious documents. J. WALPER (Ft. Worth), N. PAVONI (Zurich), D. RIGASSI (Geneva) and P.V. ZALAN (Rio de Janeiro) supplied me with original data.

J. FANSHAWE (Billings) introduced me to the foreland tectonics of the Overthrust Belt and shared with me has great geological knowledge and wisdom.

J. SCHOEFFLER (Pau), N.J. Mc MILLAN (Calgary), J. DERCOURT (U. of Paris VI), M. MATTAUER (U. of Montpellier), R. BOURROUILH (U. of Bordeaux) and D. RIGASSI (Geneva) gave me their support at critical moments of my progression.

Last but not least I wish to thank my principal advisor, critic, and corrector for both texts, ANDREW ZOLNAI (Calgary), who also provided me with his ideas, a number of field observations and numerous documents.

The responsability for the views (and possible heresies) expressed above remains nevertheless solely mine.

XII. LIST OF REFERENCES

BAARS D.L. & STEVENSON G.M. (1982)
Subtle Stratigraphic traps in Paleozoic rocks of Paradox Basin. in : (Halbouty M. ed.) AAPG. Mem. 32. pp. 131-158

BALDUCCI A. & POMMIER G. (1970)
Cambrian oil field of Hassi Messaoud, Algeria in A.A.P.G. Memoir n° 14, pp. 477-488

BARTLETT W.L., FRIEDMAN N. & LOGAN J.M (1981)
Experimental folding and faulting of rocks under confining pressure Tectonophysics, 79, pp. 255-277

BERGERAT F. (1977)
La fracturation de l'avant-pays jurassien entre les fossés de la Saône et du Rhin. Analyse et essai d'interprétation dynamique.
Rev. Géog. Phys. Géol. Dyn. Vol. XIX, fasc. 4, pp. 325-338

BOURROUILH R., RICHERT J.P. and ZOLNAI G. (1988)
The North Pyrenean Intraplate Basin : Evolution and Hydrocarbons, and : Iberia versus Europe - Effects of Continental Break-up and Round-up on hydrocarbon habitat.
A.A.P.G. Bull., V. 72, n° 8, p. 990 (Abstracts only).

BROWN W.G. (1984)
Basement involved tectonics, foreland areas
A.A.P.G. Continuing education course note series n° 26

BRUN J.P. & CHOUKROUNE P. (1983)
Normal faulting, block tilting and décollement in a stretched crust.
Tectonics V.2 n° 4 pp. 345-356

CLEMENT J.H. (1987)
Cedar Creek : A significant Paleotectonic feature of the Williston Basin
A.A.P.G. Memoir n° 41, (J.A. PETERSON Ed.), pp. 213-240

COOK D. (1983)
The Northem Franklin Mountains, N.W.T., Canada - a scale-model of the Wyoming Province.
Rocky Mtn. Assoc. Géol. :"Rocky mountain Foreland Basins and Uplifts" pp. 315-338

CRAWFORD F.D., SZELEWSKI C.E. & ALVEY G.D. (1985)
Geology and exploration in the Takutu graben of Guyana and Brazil.
Journ. of Petr. Geol. V. 8 n° 1, pp. 5-36

CROWELL J.C. (1962)
Displacement along the San Andreas Fault, California in : "Special G.S.A. Papers" n° 71

CROWELL J.D (1974)
Origin of Late Cenozoic basins in Southern California in : "Tectonics & sedimentation" (W.R. Dickinson ed.).
S.E.P.M. Spec. Publ. 22, Tulsa. pp. 190-204

DE VOTO R.H. (1980)
Pennsylvanian stratigraphy and history of Colorado - Rocky Mtn. Assoc. Geol.
Symposium pp. 71-101

FINLAYSON D.M., LEVEN J.H. & ETHRIDGE M.A. (1988)
Structural styles and bassin evolution in Eromanga Region, Eastern Australia.
A.A.P.G. Bull V. 72, n°1, pp. 33-48

FOX F.G (1985)
Structural geology of the Parry Island fold belt (Arctic Canada).
Bull. Can. Petr. Geol. V. 33, n° 3, pp. 306-340

GATEWOOD L.E. (1970)
Oklahoma City field - anatomy of a giant. in : AAPG. Memoir 14 "Geology of giant petroleum fields", (M.E. Halbouty ed.)
pp. 223-254

GELARD J.P. & RAT P.(1977)
Vues actuelles sur les grandes lignes de la structure de la Bourgogne en relation avec l'évolution post-hercynienne.
Vème réunion annuelle des Sc. de la Terre - Rennes - Abstracts p. 243

GELARD J.P. (1978)
La fracturation de la Bourgogne méridionale, essai d'interprétation et implications tectoniques régionales.
Rev. Géogr. Phys. Géol. Dyn. (2) Vol. XX. Fasc.1. pp. 5-28

GWYNN T.A. (1964)
The Cedar Creek Anticline.
3rd internatl. Williston Basin Symposium pp. 192-199

HAM W.E., DENISON R.D. and MERRITT C.A. (1964)
Basement rocks and structural evolution - Southern Oklahoma.
Okla. Geol. Surv. Bul. 95

HARDING T.P. (1974)
Petroleum traps associated with Wrench Faults.
A.A.P.G. Bull V. 58 n° 7, pp 1290-1304

HARDING T.P. & LOWELL J.D. (1979)
Structural styles, their plate tectonic habitats and hydrocarbon traps in petroleum provinces.
A.A.P.G. Bull V. 63, n° 7, pp 1016-1058

HARDING T.P. (1983)
Divergent wrench fault and negative flower structure, Andaman sca. in : (A.W. Bally ed.) "Seismic expression of structural styles".
A.A.P.G. Studies in Geology Series N°15. pp. 4.2-1 to 4.2-8

HARDING T.P., GREGORY R.F. & STEPHENS L.M. (1983)
Convergent wrench fault and positive flower structure, Ardmore Basin, Oklahoma.
Ibid. pp. 4.2-13 to 4.2-17

HENRY J. (1983)
Manuel d'analyse structurale. (Dans : Méthodes modernes de géologie de terrain, n° 2a) - Editions Technip.

HOFFMAN P.F., BELL I.R., HILDEBRAND R.S. and THORSTAD L. (1977)
Geology of the Athapuscow Aulacogen, East Arm of Great Slave Lake, District of Mackenzie.
Geol. Surv. of Canada, Rep. of A ct., 77-1/A, pp. 117-129

JOHNSON W.D. & SMITH H.R. (1964)
Geology of the Winnett - Mosby Area... Montana
U.S.G.S. Bulletin n° 1149

KINGMA J.T. (1958)
Possible origin of piercment structures, local unconformities and secondary basins in the Eastern Geosyncline, New Zealand
N.Z. Journ. of Geol. & Geoph. V.1, n° 2, pp. 269-274

LAROCHE P.J. (1983)
Appalachians of Southern Quebec seen through seismic line n° 2001 - in (A. W. Bally ed.) "Seismic expression of structural styles".
A.A.P.G Studies in Geology Series n° 15, pp. 3.2.1-7 to 3.2.1-22

LOWELL J.D. (1972)
Spitzbergen Tertiary Orogenic Belt and the Spitzbergen Fracture Zone.
Geol. Soc. Am. Bull V. 83, n° 10, pp. 3091-3102

MATTAUER M. (1980)
Les déformations des matériaux de l'écorce terrestre
Hermann Paris (ed.) - Collection Méthodes, 2è édition

MATTE Ph. (1986)
La chaîne Varisque... au Permo-Carbonifère
Bull. Soc. Geol. Fr. (8) t. II., n° 1 pp. 9-24

MEILLIEZ F. (1987)
Géologie du Massif Armoricain
Géochronique n° 21, p. 10

MOLNAR P. & TAPPONIER P. (1975)
Cenozoic tectonics of Asia : effects of a continental collision.
Science V. 189, n° 4201, pp. 419-426

MOODY J.D. & HILL M.J (1956)
Wrench-fault tectonics.
Geol. Soc. Am. Bull. V. 67, pp. 1207-1246

MOODY J.D. (1973)
Petroleum exploration aspect of wrench-fault tectonics.
A.A.P.G. Bull., V. 57, n° 3, pp. 449-476

MUEHLBERGER W.R. (1980)
The Shape of North America during the Precambrian, in : "Continental Tectonics".
Nat'l Acad. Sc. Geoph. Res. Board., p. 175-182

OKULITCH A.V., PACKARD J.J. & ZOLNAI A.I. (1986)
Evolution of the Boothia Uplift, arctic Canada.
Can. Journ. Earth Sci. V. 23, n° 3, pp. 350-358

PATTERSON J.G. (1986)
The Amer Belt ; remnant of an Aphebian foreland fold and thrust belt.
Can. J. Earth Sci. V. 23, pp. 2012-2023

PAVONI N. (1961)
Faltung durch Horizontalverschiebung.
Ecl. Geol. Helv. V. 54, n° 2, pp. 515-534

PRITCHETT R.W. (1985)
Seismic Profiles of the Western Laramie Basin-Wyoming in : "Seismic Exploration of the Rocky Mountain Region" (R.R. Gries & R.C. Dyer editors) R.M. Assoc. Geol. & Denver Geoph. Soc. pp. 225-232

RAY R.R. & KEEFER W.R (1985)
Wind River Basin, Central Wyoming - in : "Seismic Exploration of the Rocky Mountain Region" (R.R. Gries & R.C. Dyer editors)
R.M. Assoc. Geol. & Denver Geophys. Soc. pp. 201-212

READING H.G. (1980)
Characteristics and recognition of strike-slip fault systems.
Spec. Publ. Intern Assoc. Sedim. n° 4, pp. 7-26

RIEDEL W. (1929)
Zur Mechanik geologischer Brucherscheinungen
Zentralblatt Min., Geol., Pal. - Abt.B.

RIGASSI D.A. (1986)
Wrench-faults as a factor controlling petroleum occurences in West Siberia
A.A.P.G. Memoir n° 40,pp. 529-544

RODGERS D.A. (1980)
Analysis of pull-apart basin development produced by "en-echelon" strike-slip faults.
Spec. Publ. Int. Ass. Sedim. 4, pp. 27-41

ROLET J. (1984)
Graben losangiques (pull-apart) en régime de décrochement.
Le rôle des coulissements hercyniens dans l'individualisation des bassins carbonifères du Massif Armoricain.
Ann. Soc. Geol. Nord, Tome CIII, pp. 209-220

RUHLAND M. (1972)
Le rôle des décrochements dans le socle vosgien
et en bordure du Fossé Rhénan. in : "Approaches to taphrogenesis"
Procedings of Int. Rift Symp.
Karlsruhe (13-15.IV.1972), pp. 167-171

SALES J.K. (1971)
Structure of the northern margin of the Green River Basin, Wyoming.
Wyo. Geol. Assoc. Guidebook pp. 85-102

STEARNS D.W. (1971)
Mechanisms of drape folding in the Wioming Province
Wyo. Geol. Assoc. - 23rd Ann. Field Conf. - pp. 125-143

STEARNS D.W. (1978)
A.A.P.G. Structure Seminar - Notes (Nov. 22-23, Calgary) pp. 1-15

STONE D.S. (1974)
Lineaments, their role in tectonics of Central Rocky Mountains.
Wyo. Geol. Assoc. Earth Sc. Bull. V. 7, n° 4, pp. 1-11

SYLVESTER A.G. & SMITH R.R. (1976)
Tectonic transpression and basement-controlled de formation in San Andreas fault. zone, Salton trough, California.
A.A.P.G. Bull. V. 60, n° 12, pp.

TAPPONIER P. PELTZER G. & ARMIJO R. (1986)
On the mechanics of the collision between India and Asia-In : Collision Tectonics (Coward M.P. & Ries A.C. eds).
Geol.Soc.Spec. Publ. n° 19, pp. 115-157

TCHALENKO J.S. (1970)
Similarities between Shear Zones of different magnitudes.
Geol.Soc.Am. Bull. V. 81, pp. 1625-1640

THOMAS G.E. (1971)
Continental plate tectonics ; Southwest Wyoming.
Wyo.Geol.Assoc. Guidebook pp. 103

THOMAS G.E. (1974)
Lineament-block block tectonics ; Williston - Blood Creek Basin.
A.A.P.G. Bull. V. 58, n° 7, pp. 1305-1322

TUZO WILSON J. (1965)
A new class of Faults and their bearing ou Continental Drift.
Nature n° 4995, pp. 343-347

U.S.G.S (1975)
Paleotectonic investigations of the Pennsylvanian system in the U.S.
Professional Paper 853 (Mc Kee & Crosby ed.)

WALPER J.L. (1977)
Paleozoic tectonics of the Southern margin of North America
Gulf Coast Assoc. Geol. Soc. 27th ann-mtg
Austin (TEx) 26-28-X-77, pp. 230-241

WEGMAN E. (1961)
Anatomie comparée des hypothèses sur les plissements de couverture (le Jura plissé).
Bull. Géol. Inst. Univ. Uppsala, n° 40, pp. 169-182

WILCOX R.E., HARDING T.P. & SEELY D.R. (1973)
Basic Wrench Tectonics.
A.A.P.G. Bull V. 57, n° 1, pp. 79-96

WISE D.U. (1963)
An outrageous hypothesis for the tectonic pattern of the North American Cordillera.
Geol. Soc. Am. Bull. V. pp. 357-362

ZALAN P.V. (1987)
Identification of Strike-Slip faults in seismic sections.
Soc. Expl.Geoph. 57 th Ann. Mtg.
New Orleans 1987 pp. 116-118

ZOLNAI G. (1971)
Le front Nord des Pyrénées Occidentales
Histoire Structurale du Golfe de Gascogne - Technip Ed. pp. IV-5,1-9

ZOLNAI G. (1975)
Sur l'existence d'un réseau de failles de décrochement dans l'avant-pays Nord des Pyrénées Occidentales (Wrench-faults in the northern foreland of the western Pyrénées-France).
Rev. Géogr. Phys. Géol. Dyn. Vol. XVII. fasc. 3, pp. 219-238

ZOLNAI G. (1982)
Late Mississippian-Pennsylvanien Orogenic Movements of West Canadian Platform and Adjacent Areas - Their role in sedimentation and hydrocarbon accumulation.
A.A.P.G. Bull., V. 66, n° 5, pp. 646-647 (abstract only).

ZOLNAI G. (1985)
Les aulacogènes du continent Nord-Américain.
Bull. Soc. Géol. Fr. V.8, t. II, n° 5, pp. 809-818

ZOLNAI G. (1989)
Arches, lineaments, mega-shears, rifts and aulacogens of North America ; to be published in the : Bull des Centres de Recherche - ELF Aquitaine, in May 1989.

A.A.P.G.

"TECTONIQUE CONTINENTALE EN CISAILLEMENT"

Séminaire

("short course") n° 2

Animateur : Greg ZOLNAI

Dimanche 25 Septembre 1988, 08:00 - 17:00 hrs

/ Texte français /

G. ZOLNAI

ELF-Aquitaine

PAU (64018) France

(adresse présente : 11 Av. des Sayettes

64000 PAU - FRANCE)

RESUME

1) A l'intérieur des continents, la plupart des anomalies structurales telles rifts, voussures, graben rhombiques, blocs faillés surélevés voire plis et groupes de plis sont dus partiellement ou entièrement à des cisaillements.

2) A cause de la grande épaisseur et de la rigidité de la croûte continentale, des déformations véritablement compressives avec raccourcissement crustal sont extrêmement rares. Des coulissages modestes sont cependant capables de créer d'importantes structures positives et négatives et d'initier de grands systèmes de fractures. Ces derniers peuvent être maintenus en activité par de petits mouvements d'ajustement, dont la direction peut être inversée d'une phase tectonique à une autre.

3) Dans les zones de coulissage des "bassins et hauts-fonds associés" sont créés simultanément, par la réorganisation interne des masses, et non par des convergences et étirements extérieurs ou régionaux. L'intense fracturation et le morcellement causés par le cisaillement, ainsi que la friction le long des plans de failles sont à l'origine de rotations et de basculements de blocs, qui peuvent introduire des compressions et étirements secondaires ou locaux. L'inversion du cisaillement peut changer la signature des anomalies.

4) Les déformations intracontinentales sont pour la plupart concentrées autour d'anciennes sutures réactivées ou zones de faiblesse ; le rôle de l'héritage structural est donc capital. Les zones de cisaillement majeures peuvent traverser les continents de part en part sans que la quantité de mouvement qui les a engendrées disparaisse. Des anomalies secondaires de type et de taille variés peuvent apparaître tout au long de leur passage.

5) De tels systèmes de failles fondamentales -géofractures- pénètrent la croûte continentale sur toute son épaisseur. Ils peuvent être interconnectés pour former des réseaux tectoniques primaires qui vont absorber la plus grande partie des déformations intracontinentales successives. Les blocs cratoniques, ou "nuclei" qu'ils séparent, vont subir de petits mouvements d'ajustement, les uns par rapport aux autres, en forme de décrochements et/ou rotations.

6) Les plis ou systèmes de plis superficiels, que l'on rencontre à l'intérieur des continents ou le long des marges passives sont, eux aussi, le plus souvent associés à des coulissages profonds. Des groupes de plis en échelon, obliques par rapport aux failles principales sont des plis d'entraînement, tandis que les plis parallèles aux failles principales sont dus au drapage sur des panneaux du substratum éjectés, voûtés ou basculés dans les couloirs décrochants sous-jacents. Ces plis ne nécessitent donc pas de compression perpendiculaire à leur axe. De tels groupes de plis peuvent former des ceintures orogéniques "passives" importantes, loin des orogènes actifs.

7) Pendant leur dérive, les continents sont tirés plutôt que poussés. Les zones véritablement compressives sont donc limitées à des ceintures relativement étroites le long des marges actives, tandis que l'intérieur des continents est soumis à des étirements diffus, facilités par les fractures profondes. De vastes dépocentres se forment ainsi, à l'intersection des systèmes de coulissage. Des déformations orogéniques passives peuvent avoir lieu le long des marges passives, cisaillantes ou "à la traîne", où les blocs individuels disposent d'une plus grande liberté latérale.

8) Les coulissages intracontinentaux peuvent survenir à n'importe quel moment du cycle tectonique global, quoiqu'ils soient le plus souvent syn-orogéniques. Ils sont capables d'atteindre presque n'importe quel endroit à l'intérieur des continents, car les zones de faiblesse créées et entretenues par les coulissages successifs peuvent projeter les déformations sur de longues distances.

9) Il est souvent difficile d'évaluer directement les systèmes de décrochements ; leur géométrie interne spécifique peut cependant permettre la compréhension du champ de déformation et de déterminer la direction et l'ampleur du ou des déplacements successifs ; leur connaissance peut donc être très importante pour l'exploration.

10) Les systèmes en cisaillement sont très riches en hydrocarbures, car :

- de puissantes séries sédimentaires peuvent s'accumuler dans les bassins d'étirement et rifts ;

- les zones effondrées en coulissage sont propices à la création de milieux confinés (d'où : bonnes roches-mères) ;

- des pièges stratigraphiques et structuraux multiples peuvent se former dans les zones mobiles, pendant les phases de mouvements successifs ; même dans les aires continentales par ailleurs calmes ;

- les fractures et réseaux de failles verticales facilitent grandement la migration des hydrocarbures qui peuvent, grâce à ces derniers, atteindre des habitats peu profonds ;

- la fracturation peut améliorer les qualités-réservoir et accélérer la maturation grâce à des flux de chaleur élevés.

Le facteur-clé pour générer des accumulations commerciales reste cependant la présence de couvertures efficaces, capables de contre-carrer l'effet de dispersion, dû aux fentes d'étirement verticales omni-présentes.

Les mécanismes de déformation n'étant pas les mêmes, l'approche de l'exploration doit être différente selon qu'il s'agit de domaines cisaillés ou de domaines comprimés.

"TECTONIQUE CONTINENTALE EN CISAILLEMENT"

- Ebauche -

I.	MECANISMES DE BASE	1/2 HR		08:00
II.	DECROCHEMENTS ET FAILLES	1/2 HR		
III.	DECROCHEMENTS ET PLIS	1/2 HR		
			Pose	- 10:00
IV.	CRITERES D'IDENTIFICATION	1/2 HR		
V.	AMPLITUDE DES MOUVEMENTS	1/2 HR		
VI	ROLE ET INFLUENCE DES DECROCHEMENTS	1 HR		
	discussions	1/2 HR		- 12:00
	--		Déjeuner	----------
VII.	HYDROCARBURES -			
	INFLUENCE DES COULISSAGES	1 1/2 HR		13:00
VIII.	TRAVAUX PRATIQUES	1 HR		
			Pose	- 16:00
IX.	INTERPRETATION GLOBALE et			
	MECANISMES PROFONDS	1/2 HR		
X.	RECOMMANDATIONS DE TRAVAIL	1/2 HR -		- 17:00

S O M M A I R E

"Tout ce qui est simplifié est faux...
tout ce qui ne l'est pas est inutilisable."

(Paul VALERY)

AVANT-PROPOS

Les régions stables, situées à l'intérieur de chaque continent (appelées dorénavant "Intérieurs Continentaux", par opposition aux "Marges Continentales") sont des domaines vastes, stables à carac-ère cratonique et apparemment peu affectés par les déformations récentes. Ils figurent depuis longtemps parmi les régions les plus riches en hydrocarbures. Leur exploration a été basée sur des méthodes développées - et sur une philosophie acquise - dans les avant-pays orogéniques, franchement plissés et comprimés. Avec l'augmentation rapide du volume des données d'exploration et des acquis scientifiques, (et grâce à leur précision accrue) on est actuellement en mesure de tenter une approche nouvelle d'une part pour définir les caractéristiques propres à ces régions et, d'autre part pour mieux saisir les mécanismes de leur déformation. Ceci faisant on pourra également mieux comprendre le comportement des zones de transition entre ces noyaux stables et les zones orogéniques.

L'étude d'un certain nombre de structures dans les Intérieurs Continentaux, montre que des décrochements pouvaient jouer un rôle déterminant dans leur évolution. Ce type de déformation en coulissage est bien connu et étudié en détail dans certaines régions appartenant aux marges continentales (telles la Californie et la Nouvelle Zélande) qui ont subi - et subissant encore - de fortes déformations orogéniques. Les éléments de ces études ont, certes, une validité mécanique incontestable, mais la généralisation de ces mécanismes aux domaines stables, cratoniques reste toutefois douteuse. Il est donc important d'évaluer les similitudes et les différences entre les phénomènes tectoniques des deux types de domaines, et de enter de quantifier les variations de ces mécanismes.

On pourra ainsi mieux saisir la structuration des Intérieurs Continentaux et mieux évaluer leur impact sur la tectonique dite globale, ce qui n'a été fait que trop peu jusqu'ici.

I. MECANISMES DE BASE

I-A) INTRODUCTION

Le diagramme fondamental de RIEDEL (datant de 1927) donne la description et la classification des jeux de fractures principaux créés par un cisaillement parallèle. Il faut en retenir la présence de fractures et de failles synthétiques et antithétiques (Fig. I-5).

Le diagramme de HARDING définit, en plus, l'arrangement des structures associées, en compression et en extension, tels que les plis, chevauchements, failles normales, fentes ouvertes et stylolites (Fig. I-6).

Le diagramme de CROWEL résume la géométrie de base d'un bassin rhombique d'étirement ("pull-apart") ainsi que les traits secondaires que l'on rencontre dans un système de coulissage (Fig. I-7).

WILCOX et KINGMA (entre autres) ont décrit la géométrie et le comportement des zones de coulissage, pendant le processus de cisaillement (voir aussi § II) (Fig. I-8).

Les schémas mécaniques des coulissages avec relais ("off-set" ou "séparation") ont été définis par CROWEL et RODGERS (Fig. I-9).

Les "failles transcurrantes" ont été les premières à être étudiées dans le domaine des failles à rejet horizontal, sur la côte ouest de Californie (faille de San Andreas) ; l'utilisation de ce terme devrait donc être réservée à des failles et systèmes de failles qui sont très longs de plusieurs centaines voire milliers de km, continus et linéaires, et dont le rejeu horizontal est de l'ordre hecto-kilométrique (Fig. I-10).

Les "failles transformantes" océaniques ont été définies par TUZO WILSON ; l'emploi de ce terme devrait donc être limité aux domaines océaniques et évité dans des contextes continentaux, car le sens du déplacement le long de ces accidents n'est pas forcément identique à la "flèche" des failles de décrochement correspondantes sur les continents, et que les mécanismes des dorsales (rifts) océaniques sont tout-à-fait indépendants des mécanismes des Intérieurs Continentaux (Fig. I-11).

Lorsqu'une compression externe s'ajoute à un coulissage on parle de transpression. On rencontre souvent dans ce cas des structures en fleurs (appelées aussi "choux-fleurs" : "flower structures"), mais également des blocs éjectés et des rides faillées. Dans le cas inverse, lorsque la translation est associée à une extension on a affaire à une transtension, génératrice de (choux-)fleurs négatifs ainsi que de sillons, rhomb-graben, et de cuvettes faillées etc... (Fig. I-12/A-B-C)

Dans les couloirs décrochants on observe également des rotations, ainsi que des fentes ouvertes (fentes d'extension). Des systèmes de coulissage trouvent souvent leur expression en forme de failles ou groupes de failles en échelon et de plis en échelon. Des zones de cisaillement peuvent également créer des linéaments et des voussures (Fig. I-13).

Les coulissages parallèles en profondeur (donc sans compression externe) entraînent souvent la formation de plis superficiels. Ces derniers peuvent être dus directement à l'effet d'entraînement ; on parle alors de plis d'entraînement (ou "plis de torsion" : terme proposé par D. RIGASSI) ; ils peuvent également provenir d'un effet de drapage, à l'aplomb de blocs profonds éjectés, voûtés, "archés" ou basculés (plis de drapage). Dans les deux cas il s'agit de plis forcés ou "forced folds", qui ne sont causés ni par une compression régionale ni par un raccourcissement crustal (par opposition aux "free folds"-D. STEARNS) (Fig. I-16-17-18).

Des dispositifs en "parquets" sont souvent présents au niveau du socle (ou dans des terrains consolidés) ; il s'agit de familles de failles parallèles, longues, régulièrement espacées qui subissent des coulissages dans le même sens (Fig. I-17 et 18).

I-B) FIGURES

I-C) CONCLUSIONS

- Lorsque la contrainte initiale est en "cisaillement simple" (lorsqu'il existe un couple initial) on verra (Fig. I-14) :

  (a) une distribution de cassures régulière, conjuguée mais asymétrique ou non-conjuguée,

  (b) de nombreuses fentes d'extension ouvertes ("R'" de RIEDEL), parfois sigmoïdales,

  (c) des plis en échelon, (souvent individuellement courts), c'est-à-dire dans des groupes parallèles mais décalés les uns par rapport aux autres, les axes des plis formant un angle aigu avec la direction de la force primaire (THOMAS).

- Si la contrainte initiale est en compression (en "cisaillement pur") ou en extension (Fig. I-14) :

  (a) les cassures seront distribuées d'une façon symétrique (jeux conjugués),

  (b) les fractures seront pour la plupart fermées,

  (c) les plis (qui pourront être longs et légèrement arqués) formeront des groupes symétriques ou des alignements, leurs axes étant perpendiculaires par rapport à la force initiale.

- Dans les terrains durs et compétents le mécanisme de base, c'est-à-dire le coulissage des blocs, entraînera les phénomènes successifs suivants :

  . une friction (au moins en profondeur),

  . une (tendance à la) rotation des blocs "débités" par les fractures (R'),

  . des compressions secondaires ou locales, car les blocs (losangiques) vont "coincer" entre les parois du système,

  . des blocs auront tendance à monter, en se voussant et en chevauchant légèrement les mâchoires ou compartiments voisins.

  . ces "uplifts" seront relayés ou compensés par des effondrements locaux.

- Dans des cas de translation simple ou parallèle, les systèmes en damiers équilibrés ("paired uplifts and basins" - J. WALPER) assureront l'équilibre des masses (cf. plus loin, § II. et V.). Les "uplifts" ou horsts, formeront la majorité en coulissage de transpression, alors que les effondrements, bassins, "pull-aparts" caractériseront les cisaillements en transtension. Il pourra cependant toujours y avoir des blocs qui agiront à contre-sens (donc des effondrements, graben dans les systèmes transpressifs et des buttes tectoniques émergées dans les conditions d'effondrement général ou de transtention).

- Dans le cas de terrains mous et incompétents, la formation des plis de drapage et d'entraînement résulte également d'une friction interne des roches, qui leur assure leur cohérence et donc leur plasticité. La contraction locale, qui va être compensée "sur place" par une élongation, sera responsable des plis d'entraînement. Les faisceaux de synclinaux et anticlinaux ainsi créés pourront subir également des rotations, dans le courant du coulissage.

- Un bloc rehaussé (uplift) ou la présence d'un pli ne signifie pas nécessairement qu'il y a compression, dans le sens régional ("raccourcissement "ou" compression orogénique") ; un système de fractures ouvertes ou une cuvette ne prouve pas qu'une extension généralisée a eu lieu ("rifting"). L'un et l'autre peuvent être dus à un cisaillement, qui va générer son propre champ de contraintes secondaires, avec des compressions et extensions d'un ordre inférieur. Les éléments (anomalies) structuraux positifs et négatifs se présenteront dans ce cas le plus souvent "en paires". Ces structures doivent donc leur existence à une réorganisation interne des masses pris dans le processus du coulissage, plutôt qu'à un raccourcissement on à un serrage (Cf. § V ci-après).

  Cette réorganisation interne se manifestera différemment dans les terrains ou "niveaux tectoniques" différents. Les failles et les jeux de blocs en damiers caractériseront les niveaux inférieurs compétents, tandis qu'on observera des plis "forcés" dans les niveaux structuraux supérieurs, incompétents (cf. respectivement chapitres II. et III. ci-après).

ATTENTION :

Des plis véritablement compressifs et des plis "secondaires" (du type entraînement et drapage), peuvent se former d'une façon synchrone et dans des domaines extrêmement proches, par exemple dans les avant-pays orogéniques. Il est extêmement important, du point de vue de l'exploration, de savoir "qui est quoi" ?

Rien ne ressemble plus à un pli d'entraînement dans un système de plis en échelon, qu'un pli de drapage, formé à l'aplomb de la crête (tête) d'un panneau basculé dans un couloir de cisaillement. Ce qui peut induire en erreur quant à l'estimation du sens du coulissage (Fig. VIII-7).

L'un des obstacles fréquents à l'application de la tectonique par cisaillement dans les Intérieurs Continentaux est que l'on réfléchit trop souvent en termes de "cisaillement pur" (en fait en compression) et que l'on a du mal à admettre qu'un "couple" (ou cisaillement simple) soit à l'origine des mouvements ou déformations (Cf. plus loin, chapitres VI et IX).

II. DECROCHEMENTS ET SYSTEMES DE FAILLES

II-A) INTRODUCTION

Les cisaillements affectent d'une façon différente les terrains durs, consolidés et les terrains mous, meubles.

Dans les ensembles compétents, comme les zones cratoniques, les cisaillements auront comme effet principal la création de fractures (ou le renouvellement d'anciennes fractures). Dans des cas idéaux les éléments du système initial - composé de fractures ouvertes, parallèles, en échelon - vont "fusionner". Il se créera ainsi des bandes ou couloirs de coulissage où se détacheront des blocs rhomboidaux ou rectangulaires, entre deux lignes de "failles-maîtresses" ("master-faults").
- L'héritage du socle ancien interviendra cependant toujours dans ces terrains durs et consolidés qui se trouvent surtout dans les plates-formes anciennes - ou à l'intérieur des aires à un socle constitué. L'image finale sera donc le plus souvent un compromis, car le champ de contrainte nouveau ne produira pas librement ses propres anomalies ; mais il sélectionnera plutôt parmi les lignes structurales préexistantes celles qui lui seront les plus proches.

Quelques cas de terrain seront présentés dans les figures II-2 à II-8, permettant de visualiser ces systèmes de failles.

II-B) FIGURES

		Fig. n°
. Couloirs (bandes) de cisaillement et systèmes de fracture Montana central (USA)	/JOHNSON/	II-1
Bourgogne (France)	/BERGERAT/	II-2
. Zones de cisaillements majeures, linéaments	/BAARS/	II-3
. Bassins losangiques	/DE VOTO, PATTERSON/	II-4
	/HAM, WALPER/	II-5
. Blocs hauts et effondrements associés	/CROWELL/	II-6
. Touches de piano	/DE WIT/	II-7
	/G.Z./	II-8
. Systèmes de fracture du Montana Central	/U.S.G.S./	II-9

III. DECROCHEMENTS ET PLIS

III-A) INTRODUCTION

Les séries moins compétentes, appartenant aux niveaux structuraux supérieurs, répondront plus fréquemment au cisaillement en forme de plis et systèmes de fractures associés. - Les plis seront mieux développés quand un niveau "savonneux" de décollement (argiles, évaporites), sera intercalé entre les séries sédimentaires encore souples et leur substratum.

Quelques uns des cas bien connus vont être présentés dans les figures III-1 à III-9, concernant la relation des décrochements et failles profonds avec les plis superficiels.

III-B) FIGURES

			Fig. n°
. Systèmes de coulissage et plis superficiels (Ctr. Montana)		/G.Z./	III-1
. Plis d'entraînement :	Ste Suzanne (France, S-O),	/G.Z/	III-2
	Jura franco-suisse,	/PAVONI/	III-3-4
	Franklin Mountains du Nord (N.W.T.-Canada),	/COOK/	III-5
	Experimental	/ODONNE/	III-6
. Evolution des plis d'entraînement		/THOMAS/	III-7
. Plis de drapage	Cedar Creek (Montana, USA),	/CLEMENT/	III-8
	Province de Wyoming (USA).	/STEARNS/	III-9

IV. CRITERES D'IDENTIFICATION DES FAILLES DE DECROCHEMENT ET DES SYSTEMES DE COULISSAGE

IV-A) Sur CARTES (en projection horizontale) on peut reconnaître les décrochements par l'un ou plusieurs des critères suivants :

A) Les traits tectoniques et des indices (corps) géologiques, disposés perpendiculairement à la faille principale sont décalés le long de cette faille ou terminés abruptement contre elle. On peut ainsi constater des déplacements horizontaux, entre des réseaux de failles secondaires, des alignements récifaux, des systèmes de cheneaux, des changements de faciès, ainsi qu'entre des corps intrusifs et volcaniques et des filons (Fig. IV-1-2).

B) Les cisaillements produisent d'habitude des bandes longues et étroites à l'intérieur desquelles les anomalies parallèles, de types divers, se produisent "en échelon" donc d'une façon régulière et répétitive mais avec décalages obliques. Ces anomalies peuvent comprendre aussi bien des failles que des faisceaux de plis, ou conjointement les deux (Fig. IV-3).

C) Les anticlinaux (plis) sont souvent traversés de part en part par des failles parallèles orientées à peu près perpendiculairement à l'axe du pli. Ces fractures ne forment pas des systèmes symétriques ou conjugués ; elles sont souvent ouvertes et peuvent causer de petits décrochements.

D) Ces anticlinaux sont souvent déversés et parfois chevauchants, d'un côté ou des deux.

E) Les anomalies opposées, de part et d'autre de la faille principale ont souvent des signatures inverses. Elles peuvent être tronquées par la faille décrochante. Ce dispositif peut suggérer que la faille en présence n'est qu'une faille normale "en ciseau" dont le rejeu change souvent de côté ou de direction, ce qui serait inhabituel s'il s'agissait des simples failles normales (Fig. IV-4).

F) Des voussures, des panneaux faillés cunéiformes éjectés ou des blocs en forme de "bouchons" surélevés sont souvent associés aux couloirs de décrochement, pouvant en occuper toute la largeur (Fig. IV-6).

G) Des pointements volcaniques sont fréquents le long des (systèmes de) failles décrochantes tandis que les graben associés, s'il y en a, peuvent être remplis par des coulées.

H) Des "bassins et hauts-fonds associés", formant des systèmes allongés, témoignent souvent de coulissages à grande échelle. Les failles normales et inverses alternent et les blocs surélevés (ou les blocs affaissés) se touchent aux angles.

I) Dans des zones fortement érodés, les systèmes de cisaillement se présentent sous forme de groupes de failles fortement redressées voire verticales, qui renferment des blocs de provenance variée, en forme d'amigaloïdes et/ou synclinaux pincés (Cf. plus bas). Sur le terrain, ces systèmes forment des zones d'anomalies rectilignes, quelle que soit la topographie (ex. : la Faille Nord-Pyrénéenne, résultant d'une série d'événements transtensives, transpressives et compressives : BOURRHOUILH et al (1988) (Fig. IV-9).

IV-B) Sur COUPES (ou en projection "verticale") on observe souvent :

A) De brusques changements d'épaisseur, de composition lithologique, de faciès, de pendage, ou encore de caractère sismique entre les deux épontes des failles de cisaillement (Fig. IV-5).

B) Des structures en fleurs ("flower structures"), c'est-à-dire des "groupes de failles, divergeant vers le haut, les failles étant le plus souvent inverses" (HARDING et LOWELL - 1983) (Fig. I-12-B-C).

C) Des structures en "fleurs négatives" qui "consistent en de petits effondrements, qui chapeautent des groupes de failles divergeant vers le haut, les failles étant normales" (ibid.).

D) Des moutonnements ("effets de dauphin" de ZALAN - 1987) qui consistent en ondulations le long d'un plan de faille. L'inclinaison du plan de faille ne change pas, mais la direction du rejeu change, de normal en inverse, le long de cette faille (Fig. IV-8).

E) Des "effets de ruban" où un plan de faille subvertical change de pendage sur sa longueur et apparaît donc tantôt comme faille normale, tantôt comme faille inverse (Fig. IV-8).

F) Sur les profils sismiques on observe souvent des discontinuités le long des réflexions qui peuvent cependant ne pas être décalés verticalement ; ces zones brouillées peuvent être dues au caractère cataclastique, intensément broyé des décrochements (ZOLNAI-1975, ZALAN-1987) (Fig. IV-7).

G) Des blocs allochtones, erratiques, égarés dans les zones de failles, provenant souvent de niveaux plus profonds (pouvant comprendre même du matériau mantélique) qui forment des lentilles verticales, amigdaloïdales, coincés ou injectés dans le décrochement (Fig. IV-9).

H) Des failles juxtaposées, parallèles, apparaîssant dans la même section mais tantôt normales tantôt inverses, indiquent des coulissages possibles, parallèles à la direction des plans de faille.

ATTENTION :

- Les changements d'épaisseur (et de caractère) des couches, observés de part et d'autre des failles de décrochement sont souvent dus non pas à l'effet du coulissage même (décalage horizontal des couches) mais à une géométrie de pré-coulissage : des failles normales synsédimentaires ont très souvent un rejeu en décrochement, par exemple en bordure de graben.

- Lorsqu'on observe des corps instrusifs (ou volcaniques) de type et d'âge semblable, décalés le long des coulissages, la distance séparant ces corps n'indique pas nécessairement la "flèche" exacte ou réelle du mouvement. Il se peut que ces intrusions etc... aient été mises en place d'emblée avec un certain décalage, par exemple dans un croisement en baïonnette. De même des filons, alignements récifaux et même des corps gréseux chenalisants ont pu, dès leur mise en place ou déposition, décrire des "chicanes" au passage d'un accident pré-existant. Seules, la présence simultanée de plusieurs anomalies, si possible de types différents, qui se trouvent décalées de la même manière, permet d'affirmer avec un certain degré de confiance, la direction et la magnitude d'un décrochement.

IV-C) FIGURES

V. AMPLITUDE DES MOUVEMENTS

V-A) INTRODUCTION

Le concept des cisaillements intra- ou transcontinentaux bute contre la difficulté de mesurer ou d'estimer l'ampleur du mouvement horizontal effectif ayant lieu le long d'une faille de décrochement, ou nécessaire pour créer une bande coulissante.

- Le problème fondamental de l'observation directe réside dans la constitution même des bassins sédimentaires dont nous nous occupons principalement : nous possédons des marqueurs horizontaux en nombre plus que suffisant (strates) alors que les marqueurs verticaux, nécessaires pour observer des décrochements, sont extrêmement rares, sinon absents. Cette disproportion est de plusieurs ordres de grandeur car les filons, alignements récifaux, systèmes chenalisants etc. sont très rares et souvent mal datés. - C'est ce manque de preuves directes qui nous pousse vers l'étude de phénomènes secondaires, dont l'ensemble peut être groupé dans un système de "preuves de circonstance", qui permet d'aborder l'étude souvent très complexe des cisaillements.

- Lorsqu'on tente d'estimer l'amplitude des mouvements, d'autres difficultés surgissent. D'abord on associe instinctivement un déplacement important à toute ligne de cisaillement très longue (par exemple déplacement d'une centaine de kilomètres pour un système de coulissage long d'un millier de kilomètres). Ensuite, et dans le cas de bassins d'effondrement en losange ("pull-aparts"), on a le réflexe de chercher des mouvements latéraux (coulissages) équivalents à la longueur de l'un des côtés du parallélépipède (décrochement d'au-moins 50 km pour un bassin d'étirement de 50 km x 100 km). On finit par se demander comment les continents (ou blocs continentaux) ne se sont pas désintégrés complètement sous l'effet de déchirures de pareille ampleur. Et l'on préfère abandonner l'idée de cisaillement pour revenir vers un autre mécanisme plus conservatif, en extension-compression pure, pour expliquer les anomalies secondaires alignées tout au long de l'accident principal.

- Heureusement, les observations faites par les méthodes géophysiques profondes (sismique, méthodes potentielles) ont établi que les grands accidents ou "géofractures" affectent la croûte sur toute son épaisseur (remontée de la discontinuité de Mohorovicic à l'aplomb des rifts). Ceci nous oblige à prendre en considération la balance des masses de toute la croûte, au lieu d'envisager les déformations observées comme des phénomènes seulement épidermiques. Une voie est ainsi ouverte à l'estimation quantitative des mouvements horizontaux, à condition de connaître le volume des anomalies (positives ou négatives).

Dans ce qui suit on essayera, à partir de quelques schémas géométriques simples, de se former une idée sur l'amplitude des mouvements horizontaux intervenant lors des coulissages.

V-B) FIGURES

		Fig. n°
Diagrammes de bilan des masses :		
. Compression	/G.Z./	V-1
. Etirement	/G.Z./	V-2
	/CROWELL/	V-3
. Ouverture-rotation	/G.Z./	V-4
. Basculement	/G.Z./	V-5
. Extension par croûte mince ou épaisse	/G.Z./	V-6
Systèmes de coulissage :		
. Rotation de blocs	/G.Z./	V-7
. Ejection de blocs	/G.Z./	V-8
. Voussure de panneaux	/RUHLAND/	V-9
. Un "uplift" de taille, mise en place par coulissage : Monts Uinta (Utah, USA)	/G.Z./	V-10
. Perte de déport	/KINGMA/	V-11
. Déport constant	/G.Z./	V-12
. Bilan des volumes posifif	/BARTLETT/	V-13
. Balance des volumes-résumé	/G.Z./	V-14

V-C) CONCLUSIONS

Si le rôle des coulissages dans l'architecture des continents a été négligé ou sous-estimé pendant si longtemps, c'est probablement grâce à la "modestie" des mouvements horizontaux requis pour la génèse des reliefs importants. Ces faibles coulissages sont en effet très difficiles à identifier, à mesurer ou à estimer.

- Des mouvements horizontaux de faible ampleur (1 à 10 km), en compression ou en étirement, suffisent pour créer des reliefs positifs ou négatifs importants (tels que bassins, rifts, uplifts, arches, etc.) à condition que la croûte continentale soit épaisse. Ceci est en général le cas dans les "intérieurs continentaux", où la discontinuité dite "Moho" se trouve entre 30 et 50 km de profondeur (Fig. V-1-5)

- S'il s'agit du coulissage, l'effet vertical est du même ordre que le décalage horizontal ou supérieur. Les voussures, blocs éjectés et autres anomalies seront en plus plus abrupts, plus francs.

- L'épaisseur de la croûte sous-jacente devient le facteur-clef pour l'estimation (ou la quantification) des mouvements horizontaux (Fig. V-6) :

  . si la lithosphère est épaisse, des mouvements horizontaux peu importants, (de l'ordre kilométrique) en compression ou étirement suffisent pour mettre en place des anomalies verticales 5 ou 10 fois plus amples (par exemple dans le cas des coulissages présentant un relais) ;

  . si par contre la croûte est mince, un décrochement peut produire un bassin en étirement de la même longueur (ou largeur), la profondeur du bassin étant dans ce cas proche de l'épaisseur de la lithosphère et au fond du bassin on pourra avoir une croûte océanique ou de transition.

- La rigidité et l'incompressibilité des cratons auront comme conséquence que la largeur des bandes cisaillantes restera constante pendant le processus de coulissage. Comme les blocs pris dans ces couloirs de décrochement ne pourront pas écarter les "mâchoires de l'étau" (c'est-à-dire les épontes encadrant la bande cisaillante) et comme le plus souvent ils vont être en rotation à cause de l'effet d'entraînement qu'exercent ces épontes, ils finiront par se vousser et ils surgiront à cause des compressions secondaires. Ils laisseront des effondrements et des graben entre eux et le long des failles de bordure de la bande cisaillante. - Il n'est cependant pas possible d'établir de règle géométrique entre le déport horizontal des épontes et le mouvement vertical des blocs éjectés dans le couloir de cisaillement, même en théorie, car en géométrie pure il n'y a n'y frottement, ni "coincement", ni rotation ni "uplift" (Fig. V-7-8).

A partir de quelques cas connus (mais controversés) on peut cependant estimer que les deux vecteurs sont du même ordre de grandeur, le déplacement horizontal étant cependant 2-3 fois supérieur au mouvement vertical d'ascenseur des noyaux éjectés (exemple : Monts Uinta, Utah USA). Dans le cas des relais (off-sets) dans un système de coulissage les voussures qui se produiront seront du même ordre que le coulissage/ compression, mais 2-3 fois supérieures à ce dernier (Fig. V-10).

- Il peut y avoir dans le contexte de l'intérieur des continents, des anomalies dues à une compression ou à un étirement "pur" (perpendiculaire à l'axe de l'anomalie ou structure) ; la plupart des déformations seront cependant causées par des coulissages où les anomalies "positives" et "négatives" seront causées par une réorganisation interne des masses affectées par le coulissage, (par des transferts de masse) et non pas par un raccourcissement ou une élongation régionaux (Fig. V-9).

- Lorsque le socle est épais et rigide, (c'est-à-dire incompressible), la même quantité de mouvement latéral peut être transmise sur de longues distances, sans perte de la "flèche". Le déport peut néanmoins être réduit ou disparaître si (et là où) des accidents latéraux en compression et extension joignent (ou se détachent de) l'accident principal (Fig. V-11-12).

- Les plis associés aux coulissages ne sont pas générés par une compression proprement dite non plus, et les axes de pli ne sont pas perpendiculaires mais obliques par rapport aux directions principales de la contrainte. La réorganisation interne des masses se produira dans ce cas à l'intérieur de chaque couche, par des amincissements/ épaississements directionnels. Ce jeu de transfert des masses déjà très difficile à observer directement sur le terrain, échappe complètement aux méthodes de subsurface (forage) et à la sismique, à cause de la difficulté de mesurer les variations de l'épaisseur réelle des couches, tout au long d'une structure donnée.

Les expériences analogiques de F. ODONNE tendent à démontrer que les plis d'entraînement apparaissent dès le début du coulissage du "socle" et qu'ils changent très vite d'orientation, avec l'amplification du décrochement, en devenant de plus en plus parallèles avec la faille-maîtresse. Les plis sont déversés vers l'extérieur, à partir de l'axe du coulissage et l'on voit apparaître, en coupe, un arrangement en "bouquet de fleurs" (Fig. III-6).

- Le "bilan des volumes" est le plus souvent positif à l'issue d'un coulissage, car :

  a) Le cisaillement introduit de la porosité de fracture et peut-être de matrice, causant la "boursoufflure" des bandes cisaillantes (expériences de BARTLETT et al.) - Fig. V-13.

  b) Lorsque les bassins d'étirement sont remplis de détritus érosionnels, dérivés des blocs surélevés voisins, ces sédiments meubles pourront remplir à ras-bord les bassins et l'excédent créera des "ponts" en forme de cônes d'éjection entre les "uplifts". Des blocs isolés deviendront ainsi des rides voire des chaînes continues. (L'excédent de volume proviendra du fait que la densité des sédiments sera bien inférieure à la densité des roches érodées beaucoup plus anciennes : d = 1,5 ± contre 2,7 à 3).

  c) Ce bilan positif pourra être encore accentué par des facteurs externes, tels intrusions et volcanisme, facilités par les fractures verticales omni-présentes. Les masses plutoniques légères (granites) augmenteront ainsi la "flottabilité" des rides issues de coulissages et leur assureront une montée isostatique ; ces arches pourront donc rester en relief pendant de longues périodes, tandis que leur érosion continue sera étalée sur des centaines de millions d'années.

- La loi fondamentale pour l'application des mécanismes de décrochement sur les Intérieurs Continentaux reste cependant :

"Libérez-vous du Syndrome de San Andreas".

VI. ROLE ET INFLUENCE DES CISAILLEMENTS SUR L'ARCHITECTURE CONTINENTALE

VI-A) INTRODUCTION

La majorité des structures qui se trouvent à l'intérieur des continents, et donc bon nombre de gîsements d'hydrocarbures, remontent au moins en partie à des déformations liées à un coulissage.
Ces décrochements peuvent affecter tout segment du réseau structural hérité, mais ils seront concentrés le plus souvent en d'étroites zones de faiblesse. Leur interdigitation forme un réseau de fracture primaire qui sépare les blocs cratoniques individuels ("nuclei") en les contournant ou en zig-zagant entre eux. Notez que ces blocs individuels sont souvent plus-ou-moins arrondis et de taille assez homogène de 5-600 km en diamètre.

Ces zones de faiblesse sont composées par une suite d'éléments tectoniques divers mis bout-à-bout tels des arches, linéaments, géofractures, "uplifts" allongés, rifts et aulacogènes, ainsi que des lignes de flexures et des effondrements ou des blocs hauts isolés.
- Ces structures diverses peuvent canaliser la déformation intracontinentale entre les marges opposées actives et passives.

Le lien génétique de ces lignes de faiblesse se ramène généralement à des coulissages. Des phases de réactivation successives à mouvements opposés témoignent de la mobilité endurante de ces lignes de faiblesse.

On peut ainsi constater des reversement tels (Fig. IV-1) :

. des "pull-aparts" à des haut-fonds faillés (uplifts),

. des failles normales (rifts, sillons) à des failles de décrochement et vice-versa,

. des arches anciennes à des rifts ou à des effondrements d'ensemble.

. Des massifs importants (rides, "ranges") peuvent surgir de l'intérieur des effondrements (cas fréquent dans les aulacogènes), ainsi que

. d'anciens systèmes de charnières peuvent devenir des rifts ou, au contraire, se transformer en barrières tectoniques (en particulier près des périphéries des continents).

. Des réseaux de failles anciennes ou faisceaux de failles "en trompette" peuvent être réactivés par de faibles étirements et/ou cisaillements pour devenir des damiers de blocs basculés, des parquets tectoniques ou des touches de piano.

La plupart - sinon la totalité - des grandes cuvettes intracontinentales, dépocentres et dépressions cachent en profondeur un rift ou système de rifts (aulacogène) et elles se trouvent souvent à l'aplomb d'une jonction de tels rifts : ce sont les "fontanelles".

La persistance de leur subsidence quels que soient les changements des régimes de contraintes, peut s'expliquer par l'existence, à ces "carrefours crustaux" d'une branche prête à céder à quelconque étirement crustal. Des effondrements similaires mais plus petits et plus profonds, de véritables "entonnoirs" se forment dans les avant-pays orogéniques, aux croisements des fronts de chevauchement supra-crustaux avec des systèmes de failles intracrustaux. (Fig. VI-2).

Les blocs continentaux (nuclei) peuvent se scinder sous l'effet des rotations et coulissages en transtension ; de nouveaux grabens et rifts intracratoniques peuvent ainsi apparaître. Ces derniers s'ouvrent très souvent vers la bordure du craton ou du bloc cratonique. (*)
(Cf. p. 24)

Des phénomènes similaires existent au long des marges passives ; leur ampleur est cependant plus grande, probablement à cause de leur mobilité crustale plus grande, liée à l'amincissement de la croûte et à un plus grand degré de liberté de mouvement latéral. Les structures les plus typiques au long de la marge "traînante" sont les failles normales souvent listriques, au long des marges cisaillantes des failles décrochantes (ex. : la bordure arctique du Continent Nord-Américain, Fig. VI-10).

Les grabens péri-cratoniques, formés puis déformés en étirement et cisaillement, constituent un vaste domaine méconnu. Leur remblai sédimentaire est très épais et varié, avec des évaporites indiquant une séparation du domaine océanique. (Voici quelque temps on les aurait appelés des "miogeosynclinaux", par opposition aux "eu-geosynclinaux" que l'on considère à l'heure actuelle comme des prismes sédimentaires, progradant vers le grand large et rebaptisés en "eu-geokline" par certains auteurs américains récents). Ces effondrements périphériques peuvent facilement devenir des zones de cisaillement ou bien des parties d'orogènes, en cas d'inversion complète de la polarité tectonique (Fig. VI-4).

La traîne ou la rotation des blocs continentaux de taille considérable détachés des marges passives peut introduire des déformations très importantes, en coulissage et en transpression. Des puissants systèmes de décrochement à compressions secondaires sont capables de générer d'importants reliefs ("oro-genesis") et ne manqueront pas de manifestations volcaniques, ainsi que de plis du type drapage et entraînement amplifiés par l'halocinèse. L'absence de métamorphisme généralisé, de vastes intrusions, de masses ophiolitiques et de volcanisme de type "arc" caractéristiques des vrais "orogènes" indique que ces zones de déformations sont les expressions d'une "tectonique passive" n'impliquant ni collision, ni subduction ni raccourcissements crustaux importants (Fig. VI-10-11).

Divers coulissages existent le long des marges actives et à l'intérieur des zones orogéniques. Bien que leur analyse n'est pas notre propos, on ne peut pas les négliger totalement car certains cisaillements des marges orogéniques rejoignent ceux des régions cratoniques tandis que d'autres peuvent devenir par la suite des coulissages intracontinentaux (Fig. VI-6-9).

- Les orogénèses de collision entre Continent et Océan montrent une majorité de failles transcurrantes le long de la bordure continentale (par exemple Californie-Alaska, Indonésie, Nouvelle Zélande). Ces systèmes, chacun de plus de mille kilomètres en longueur, ont un déplacement effectif hecto-kilométrique. Des complexes de failles similaires, situés à l'intérieur des zones orogéniques sont parallèles à la marge continentale, mais leur déplacement est nettement inférieur, de l'ordre déca-kilométrique (exemple : failles de "Tintina" au Canada, "Kaltag" en Alaska). Le sens de leur déplacement est identique à celles situées le long des côtes.

  Des systèmes de cisaillement "transverses" peuvent recouper les zones orogéniques sur toute leur largeur, tel les linéaments de Texas et Lewis-and-Clark de la Cordillère nord-américaine. Le déplacement réel le long de ces coulissages est plus petit encore, de 1 à 10 Km, (exceptionnellement plusieurs dizaines de km).

- Les Orogènes ou zones de collision entre Continents n'auront pas de cisaillements ("transcurrentes") externes, actuels ou visibles. Les cisaillements intra-montagneux sont au contraire bien développés plus à l'intérieur et sont post-orogéniques ; ils déterminent l'ouverture des "bassins-successeurs" (successor basins). Ces effondrements sont d'âge Carbonifère, dans les Appalaches du Nord, triassiques dans les Appalaches du sud.

- Les zones de convergence intra-continentales (Orogènes Intra-cratoniques) ont, elles aussi, des systèmes de cisaillement parallèles à l'axe de la chaîne ainsi que des failles de décrochement perpendiculaires. Une ligne orogénique typiquement intracratonique qui a connu plusieurs périodes de convergeance (hercynien, alpin), la faille nord-pyrénéenne longe la bordure nord du bâti alpin, dont l'axe est parallèle à l'ancienne structure (hercynienne) : N-110°. Son activité en extension (trans-tension) pendant le Mésozoïque, en particulier pendant le Crétacé, a fixé la bordure sud des sillons de flysch. Un système de failles décrochantes perpendiculaires d'orientation N-S, tardi- et post-orogéniques ont également pénétré l'avant pays (ZOLNAI - 1971, 1975).

- Les exemples les mieux connus des zones de plissement en avant-pays orogéniques sont le Jura plissé franco-suisse, les Franklin Moutains dans les Territoires du Nord-Ouest du Canada et possiblement l'Atlas Saharien en Algérie. La bordure du craton donnant vers la zone orogénique principale - a été dans les deux cas - remobilisée en coulissages, selon des "lattes" ou "parquets" parallèles. Les plis d'avant-pays ne sont pas dus directement au serrage orogénique, mais ils correspondent à des plis d'entraînement de la couverture sédimentaire, au-dessus de coulissages profonds. Il faut y noter la présence de séries évaporitiques, intercalées entre le socle et la couverture (Cambrien supérieur au Canada, Trias dans le Jura).

- La remobilisation, en coulissage, d'aulacogènes ("mega-rifts" anciens ou sillons intracratoniques) sont fonction des événements orogéniques qui affectent les bordures continentales sur lesquelles débouchent ces aulacogènes. Les Monts Ougarta dans le Sahara nord-occidental d'âge hercynien, le système "East Arm" du grand Lac des Esclaves au Canada (Protérozoïque inférieur à moyen) et peut-être les plis du Bassin d'Amadeus en Australie Centrale (Paléozoïque), montrent des groupes de plis qui sont induits par des coulissages le long des faisceaux de failles propres aux aulacogènes, ainsi que des plis de drapage et d'entraînement, et des nappes gravitaires. L'ampleur des coulissages peut y être de l'ordre kilométrique à déca-kilométrique.

- Les figures VI-12 à 22 montrent quelques cas typiques de zones de cisaillements "composites".

(*) Cf. Bassin de Paradox, dans le Plateau de Colorado et le Bassin de Parentis, dans le Bloc d'Aquitaine.

VI-B) FIGURES

Fig. n°

. Le système des "Rocheuses Ancestrales" (Paléozoïque supérieur) (W.USA)	/G.Z./	VI-19
	/U.S.G.S./	VI-19/B
Rocheuses "Ancestrales" et "Laramiennes"		VI-19/C-D-E
. Bretagne (France)		VI-20
. L'effondrement central d'Australie	/G.Z./	VI-21
. Indonésie - Malaisie	/G.Z./	VI-22
. L'Est de la France : Bourgogne, Jura Région Rhin-Rhône	/RUHLAND, GELARD, BERGERAT, G.Z./	VI-23

4) Conséquences - Effets secondaires :

. Diapirisme - Halocinèse (Sillon Franklinien, Canada)	/FOX/	VI-24
. Image gravimétrique (U.S.A.)	/G.Z./	VI-25

VI-C) CONCLUSIONS

- Les récentes investigations de sismique profonde (COCORP, LITHOPROBE, ECORS etc.) ont confirmé que les rifts, aulacogènes et autres sillons superficiels correspondent en profondeur à des remontées de la discontinuité de Mohorovicic. Ceci a déjà été soupçonné auparavant, grâce aux méthodes potentielles (gravité, magnétisme) qui affichent des anomalies positives à l'aplomb des effondrements, (malgré que ceux-ci sont le plus souvent remplis de sédiments légers et récents). L'amincissements de la lithosphère peut s'interpréter comme résultat d'un étirement crustal (Fig. VI-25).

- La plupart des systèmes de rifts et de cuvettes faillées se trouvent perpendiculaires à des ceintures orogéniques, qu'ils recoupent le plus souvent à des points d'inflexion de la marge continentale. Le moment de leur formation (rifting initial) correspond toujours dans le temps à des périodes orogéniques. Des rifts récents, tels le Rift Est-Africain (comprenant la Mer Rouge et la Mer morte) et le Rift Ouest-Européen (Mer du Nord - Rhin - Rhône - Sardaigne) sont accompagnés par des coulissages. On est en droit d'admettre que ces coulissages (sénestres aussi bien en Afrique orientale qu'en Europe occidentale) ont initié plutôt que suivi le rifting, et qu'ils ont été en même temps responsables des phénomènes accompagnant la rupture, tel le volcanisme et le rehaussement des "lèvres" des rifts.

  Une fois en place, ces systèmes resteront mobiles sous les variations successives du champ de contraintes, et pourront ainsi devenir des aulacogènes (à phases d'étirement successives) ou être transformés, inversés en arches, "uplifts" (à phases de transpression) etc. Une grande partie de ces déformations ultérieures seront accompagnées de mouvement en coulissage.

- Les "Arches", tels le "Transcontinental Arch" (E.U.) et le "Peace River Arch" (CAN.) resteront longtemps en position haute (plusieurs centaines de m.a.), grâce vraisemblablement à des racines, qui correspondent à des enfoncements du "Moho" (anomalies gravimétriques diffuses négatives), qu'ils ont pu acquérir pendant des phases de compression, probablement en forme de transpression. En effet, beaucoup d'Arches et d'Uplifts s'alignent à des systèmes de coulissages majeurs connus plus à l'intérieur du continent (Tathlina Arch en face de l'East Arm System/Great Slave Lake, au Canada, Les Monts Uinta dans la prolongation du Central Colorado Trough dans l'Ouest des E.U. etc.).

- Le "rifting" de certaines de ces arches (par exemple le "Peace River Arch" pendant le Pennsylvanien) peut s'expliquer par des étirements continentaux, sans faire appel à un "doming" thermal survenant juste avant la rupture des rifts. Le rehaussement des bordures de rifts, s'il existe, n'a donc pas nécessairement été précédé par des dômes.

- Les données actuellement à notre disposition nous permettent de reconstituer les grandes lignes tectoniques transcontinentales, en mettant bout-à-bout les tronçons d'anomalies, de style et de signature parfois différents ou opposés. La plupart ont subi des coulissages directs mais tous peuvent être mieux interprétés avec une part de coulissage dans leur genèse.

- De tels cisaillements peuvent être décelés dans tous les continents ; ils font partie de l'héritage structural essentiel des plaques continentales et c'est par eux que peuvent se propager les déformations, depuis les marges actives, à travers l'intérieur des continents jusqu'aux marges passives. Les petits mouvements d'ajustement survenant le long de ces systèmes peuvent nous fournir une explication pour les anomalies de deuxième grandeur que l'on rencontre dans les zones cratoniques et qui sont très souvent situées le long de ces cisaillements ("méga-cisaillements" ou qui en sont des dépendances latérales).
  On peut ainsi mieux comprendre la présence d'"uplifts" ou de plis isolés, les systèmes en horsts-et-graben et les bassins d'étirement, mais également les dépocentres et certains faisceaux de plis.

- Chaque nouvelle phase de déformation peut ajouter de nouveaux éléments au réseau tectonique de base, le plus souvent en forme de linéaments (systèmes de fractures) ou de rifts. Certains éléments ainsi que des systèmes entiers peuvent en même temps demeurer inactifs, (dormants) pendant plusieurs diastrophismes successifs.

- Les divers types d'anomalies qui composent les systèmes transcontinentaux à un moment donné, peuvent également se succéder dans le temps, au cours de l'évolution de l'un des segments du dit système.

- De petits mouvements horizontaux (en translation, transtension ou transpression) peuvent déclencher des jeux d'anomalies répétés sur de longues distances, sans atténuation de leur mouvement lors de la création de chacune des anomalies, en milieu crustal intracontinental rigide (contrairement au milieu supracrustal souple, "marginal", où chaque pli absorbe une partie de la compression). Un long système de coulissage ne signifie pas, de même un grand déplacement horizontal le long de sa trace. Le décalage dans un système de coulissage peut être le résultat de plusieurs phases tectoniques successives, dont le sens pouvait changer dans le temps.

- Des failles apparemment néoformées correspondent très souvent à des accidents très anciens en profondeur. Ainsi des relais ou off-sets constatés dans des faisceaux ou alignements de plis récents peuvent, eux aussi, avoir leurs racines dans le socle. Les décalages observés entre plis superficiels et parallèles peuvent ne pas être dus à des coulissages post-plisssement, mais à un dispositif ancien dans le socle et réactivé tel quel lors des plissements récents (par exemple en forme d'écailles de fond). Les plis ne sont dans ces cas, que l'expression superficielle, en drapage, de failles inverses en profondeur. (Exemple : la province de Wyoming, ancien craton archéen morcelé, repris lors de la phase dite Laramienne - Paléocène).

- Des systèmes de failles transcurrentes, ayant pris naissance le long des marges actives et représentant des coulissages importants (hectokilomètres) peuvent, après collision (orogenèse) devenir des failles intra-continentales, ("sutures"). Ils peuvent préserver un certain degré de mobilité et devenir des zones de coulissages à mouvements ne dépassant pas quelques dizaines de kilomètres. Encore plus tard ils pourront être transformés en "rifts" grâce à de nouveaux cisaillements en transtension, où le mouvement ne dépasse pas quelques kilomètres. Tout ceci est hypothétique mais il y a des indications permettant de dire que les "zones de faiblesse", tout en restant mobiles, se sclérosent progressivement avec l'âge.

- La relation avec les complexes orogéniques est la facette la plus complexe et la moins étudiée de la dynamique des grands systèmes décrochants. Certains des "linéaments" du bâti nord-américain (Texas lineament", "Lewis-and-Clak lineament") ne s'arrêtent pas à l'aplomb du front de chevauchement, mais continuent à travers la ceinture plissée, pour se perdre dans le domaine océanique. D'une façon similaire, certains des accidents fondamentaux, connus en surface et en géophysique, dans les Pyrénées, traversent de part en part la chaîne.

  Le style tectonique des chaînes plissées change souvent au passage de ces grands axes transverses, et le front de chevauchement est infléchi ou décalé. Des rides océaniques peuvent se développer dans leur continuation vers le large. Même l'emplacement des "terrane" semble être influencé par le passage de ces linéaments.

VII. HYDROCARBURES - INFLUENCE DES DECROCHEMENTS SUR LEUR GENERATION ET HABITAT

VII-A) FACTEURS FAVORABLES ET DEFAVORABLES

Facteurs Favorables :

A) Les lents cisaillements synsédimentaires peuvent contribuer à créer une paléogéographie complexe, voire unique comme :

. des bassins d'étirement lozangiques fermés ("pull-aparts"),

. des rentrants de la ligne de côte et des bras de mer, le long d'anciens faisceaux de failles, des sillons, des rifts et d'aulacogènes ouverts sur une marge active,

. des structures également ouvertes vers la marge continentale en touches de piano,

. des damiers composés de blocs rectangulaires et basculés.

La plupart de ces scénarios peuvent comprendre des milieux paléogéographiques confinés et abondants en matière organique algaire, voire des milieux évaporitiques. Les faibles taux de sédimentation ainsi que le faible influx clastique en début de chaque cycle de subsidence favorise le dépôt de roche-mères riches et concentrées. Ensuite une sédimentation fine clastique et bioclastique peut déboucher sur des évaporites, couvertures efficaces et favorisant la dolomitisation des carbonates, lorsqu'il y en a. Les structures en marches d'escalier en bordure de graben facilitent aussi la colonisation récifale, qui donnera une roche réservoir juxtaposée aux roches-mères. S'il existe un chemin de migration, le long des failles ou systèmes de fractures par exemple, on a un milieu idéal de piégeage pour des hydrocarbures (Fig. VII-1-2).

(Notons au passage que ces mêmes configurations structurales et sédimentaires peuvent également donner lieu à des concentrations minérales : mine de Lady Loretta en Australie Centrale).

B) Les rapides cisaillements syn-orogéniques à taux de sédimentation très élevés, seront associés à de légers étirements (transtension). Les bassins ainsi créés (pull-aparts, rifts, aulacogènes) peuvent devenir des pièges à sédiments à taux d'accumulation élevé, dont l'épaisseur est souvent difficile à estimer (exemple : Fitzroy trough en Australie occidentale, bassin d'Anadarko dans le S.W des Etats-Unis).

C) Les coulissages le long d'anciennes lignes structurales seront prêtes à céder dès le début des déformations et à créer des habitats précoces comme des hauts-fonds synchrones ("synchronous highs" de SCHOLTEN) prêts à recevoir les hydrocarbures dès leur naissance.

D) Les déformations syn-tectoniques peuvent être transmises le long des systèmes de cisaillement, loin des zones orogéniques proprement dites, vers l'intérieur des cratons. Des habitats structuraux peuvent exister loin des sites classiques, à des endroits surprenants, "au milieu de nulle part". Ces derniers peuvent être des arches aussi bien que des dômes (Fig. VII-3 à 5).

E) Comme les accidents décrochants ont souvent plusieurs périodes de vie, différents habitats peuvent être superposés stratigraphiquement. Le plan principal des déformations liées aux coulissages est, en outre, vertical et, par conséquent, les failles coupent à travers toute la colonne sédimentaire, résultant en la présence de déformations secondaires superposées (plis d'entraînements superposés, zones de porosité empilées, etc.) (Fig. VII-6).

F) Des légers décrochements peuvent suffir pour déclencher et du volcanisme (laccolithes) et du diapirisme (halo- ou argilo-cinèse), tous générateurs d'habitats d'hydrocarbures.

G) Les fractures d'étirements peuvent être présentes partout même si les déformations sont du type transpression ; elles peuvent :

. ouvrir des cheminées de migration, reliant les roche-mères profondes avec les réservoirs et pièges peu profonds, pour former des gisements près de la surface, et néanmoins remplis d'huiles légères (matures) (Fig. VI-6-7) ;

. augmenter le flux de chaleur et accélérer la maturation, juste après la sédimentation des roches ;

. améliorer la porosité des réservoirs, par fracturation, ou même des séries argileuses normalement "couvertures" en réservoirs ; il peut ainsi se former des habitats riches en hydrocarbures au fond des synclinaux ou sillons, à l'aplomb des accidents cisaillants.

Facteurs défavorables

A) Les fractures d'étirement et fentes ouvertes représentent un danger très important, voire capital. Les hydrocarbures peuvent ainsi s'échapper dès leur formation, ou à n'importe lequel des épisodes tectoniques ultérieurs (Fig. VII-8).

B) Les zones de faiblesses anciennes peuvent être préférentiellement et profondément érodées, et perdre les sites sédimentaires favorables mentionnés ci-dessus.

C) Le flux de chaleur accru peut causer une sur-maturation (transformation des huiles en gaz) à des profondeurs où normalement on devrait encore être dans la fenêtre à huile.

D) Les masses sédimentaires, dérivées des horsts (uplifts) voisins actifs et déposées presqu'immédiatement dans les grabens à forte subsidence, sont souvent chaotiques ; les lentilles sédimentaires de petite taille, les corps gréseux immatures (pas triés), la matière organique surtout ligneuse (favorable au gaz). Le tectonisme synsédimentaire produisant de nombreuses structures de petite taille, l'habitat sera fréquemment dispersé.

VII-B) FIGURES

VII-C) CONCLUSIONS

Le facteur essentiel pour la formation d'accumulations commerciales d'hydrocarbures dans les zones dominées par la tectonique cisaillante, est la présence de couvertures, assez épaisses et efficaces pour empêcher la dismigration de l'huile. On doit être particulièrement attentif à ce que les objectifs à forer ne soient pas affectés de chemins de fuite (failles remontant jusqu'à la surface). En contre-partie on pourra trouver des gisements d'huile peu à très peu profonds et remplis d'huiles légères ("matures") et presque toujours dépourvus de gaz.

En règle générale : les cisaillements anciens engendrant les conditions propices : confinement, structures précoces, etc. sont favorables, tandis que les cisaillements récents sont le plus souvent défavorables, car ils contibueront à la dispersion des hydrocarbures voire même à la destruction des habitats déjà formés.

Il ne faut pas, par contre, négliger les pièges de petite taille qui peuvent cacher des empilements d'habitats et être remplis à ras-bord avec des hydrocarbures légers - des cibles difficiles mais à haut rendement.

Dans un même domaine, la poursuite de l'exploration appelle à des techniques différentes en régime de cisaillement qu'en régime de compression. En régime de coulissage, les structures seront en échelon et ceci influencera l'implantation des "step-outs".

L'étude des mécanismes et du déroulement de la maturation-migration devient extrêmement complexe dès que l'on tient compte des effets de coulissage.

VIII. EXERCICES

1) "PLI du PANTALON"

(Comment se rassurer sur l'orientation des plis d'entraînement par rapport à la direction du colissage - discrètement !...).

2) FRACTURES et PLIS - Expérience :

A) coulissages synthétiques
B) coulissages antithétiques

. Prenez un "cutter" et fendez les feuilles selon les lignes obliques, entre les deux épontes du couloir de cisaillement ;

. Observez le déplacement des demi-points lorsqu'on applique un décrochement suivant les flèches ;

. Tâchez d'observer l'apparition et le mouvement des plis, formés simultanément.

3) "QUE PEUT CACHER UN SIMPLE PLI ?"

Dessinez différentes structurations profondes, pouvant être à l'origine du même (ou de similaires) pli(s) surperficiel(s).

Solutions : VIII-3/B-C.

4) "Qu'est donc advenu au BILAN DES VOLUMES ?"

. Comment se fait-il qu'on ne rencontre que des blocs éjectés dans le système de coulissage "a" et,

. Que des effondrements dans le système "b" ?

Solutions : VIII-4/B-C.

5) AMINCISSEMENT de la croûte par ETIREMENT

. Combien de coulissage est-il nécessaire pour produire un bassin d'effondrement (pull-apart) de 20 km x 20 km x 10 km, en supposant que l'épaisseur de la croûte soit de 20 km ? de 30 km ? ou de 40 km ?

. Utilisez l'espace en bas de la page pour dessiner quelques coupes de bassin davantage "géologiques".

Feuilles de brouillons : VIII-5/B
Solutions : VIII-5/C

6) "Jusqu'où peut-on (ou ne doit-on pas) ALLER TROP LOIN ?

Combien de raccourcissement crustal faut-il envisager pour créer :

. un pli de drapage ("a")
. un pli (système de plis) d'entraînement ("b") ?

Solutions : VIII-6/B

7) DRAPAGE OU ENTRAINEMENT ?

En se basant sur le cas des plis d'entraînement en "a", devinez la direction du coulissage en "b". Y a-t-il d'autres possibilités ?

Solutions : VIII-7/B

8) BASSIN LOSANGIQUE

. Quels sont les champs de cortrainte pouvant créer ce bassin "pull-apart" ?

. Quel champ de contrainte peut créer un système de failles compressives avec des directions similaires ?

Solutions : VIII-8/B

9) Exercice des "CISEAUX et du RUBAN ADHESIF"

. Prenez les figures VI-19/C-D-E (ou préférablement faites en des photocopies) et,

. Assemblez-les pour former un TRIPTYIQUE, puis,

. Etudiez le comportement des différents groupes de failles, pendant les phases tectoniques successives (Rocheuses "Ancestrales" et "Laramiennes").

. Pour une image plus large, référez-vous aux figures VI-6-7-8-9.

. L'évolution de cette région démontre bien pourquoi "la géologie structurale est une discipline à quatre dimensions" (D. STONE, 1974).

VIII. EXERCICES

1) "PLI du PANTALON"

2) FRACTURES et PLIS - Expérience :

3) "QUE PEUT CACHER UN SIMPLE PLI ?"

4) "Qu'est donc advenu au BILAN DES VOLUMES ?"

5) AMINCISSEMENT de la croûte par ETIREMENT

6) "Jusqu'où peut-on (ou ne doit-on pas) ALLER TROP LOIN ?

7) DRAPAGE OU ENTRAINEMENT ?

8) BASSIN LOSANGIQUE

9) Exercice des "CISEAUX et du RUBAN ADHESIF"

IX. INTERPRETATION GLOBALE ET MECANISMES PROFONDS

. Les continents ne sont pas des blocs monolithiques et inertes, uniformes et inchangeables, mais plutôt des ensembles composés de pièces de mosaïque, séparées par des réseaux de couloirs mobiles qui forment un réseau tectonique de base, dont les éléments sont liés entre eux. Ces derniers transmettent les déformations à travers l'intérieur des continents et ils adsorbent la plus grande partie des déformations affectant ces régions. Les structures apparemment isolées en milieu cratoniques se trouvent en fait le plus souvent le long de tels systèmes intracontinentaux (Fig. IX-1).

Tandis que ce réseau continental amortit les contraintes externes, de nouveaux éléments peuvent venir s'ajouter à ceux déjà en place, le plus fréquemment en forme de linéaments (géofractures) ou de rift-graben qui sont générés, les uns et les autres, sous l'influence ou l'effet de cisaillements. Les "nuclei" (ou mini-plaques) de la mosaïque continentale ne sont pas invariables non plus ; ils ont leur structuration intérieure spécifique et peuvent être scindés, fragmentés à leur tour.

. Les déformations à l'intérieur des Continents ne sont pas dus à des compressions orogéniques transcontinentales ("coast-to-coast"), mais plutôt de petits ajustements survenant entre les blocs (nuclei, mini-plaques) voisins plus ou moins liés. Les raccourcissements crustaux importants restent limités à des bandes relativement étroites, le long des marges actives directement exposées aux collisions. - Des blocs continentaux situés près des marges cisaillantes ou à la traîne subissent des rotations et peuvent éventuellement se détacher de la masse continentale, en cisaillement ou au moins avec une composante de coulissage (Fig. IX-2-3).

. Le moteur de la dérive continentale est situé en-dessous de la croûte, dans l'Astenosphère et il doit être propre à chaque continent. C'est une force qui agit comme une "traction avant" ; comme si les cellules de convection étaient plus fortes, davantage alignées et/ou synchronisées le long de la marge active. Il en résulte que (Fig. IX-2-4) :

a) les continents ne sont pas "poussés" mais "entraînés" par leur base (ou tirés) et donc,

b) la plus grande partie des continents se trouve, la plupart du temps, en extension ou étirement. Il en résulte la remarquable continuité dans le temps et la constance dans l'espace des dépocentres à l'intérieur des continents.

(Opposé à l'activité des marges, ce fait remarquable avait pu inspirer dans le passé l'image inerte des intérieurs cratoniques aussi bien que suggérer l'immobilisme des continents car les dépocentres ont été associés à des processus profonds "verticaux" que l'on avait du mal à déplacer avec la dérive).

c) Les forces d'entraînement agiront différemment sur des blocs à immersion différente (en fonction de leur épaisseur). Il en résultera des coulissages ou "rifting" dès le début de la dérive des continents (Fig. IX-5).

d) Des déformations intra-cratoniques peuvent survenir à n'importe quel moment du cycle tectonique "global", depuis le "rifting" et le "break-up", pendant la dérive et jusqu'à la collision. Elles sont néanmoins très souvent contemporaines avec les diastrophismes majeurs des marges actives en collision ou orogenèse. Plus on s'éloigne de ces zones actives orogéniques, plus le synchronisme devient flou.

e) Les systèmes de cisaillement peuvent changer le sens de leur mouvement, durant le même cycle global (par exemple de transtension senestre en dérive à transpression dextre en collision). - Deux segments voisins d'un même domaine continental peuvent subir, simultanément, un étirement et une compression respectivement (Fig. IX-5).

. Un continent n'aura réellement qu'une seule marge active par cycle, et bien qu' il apparaîtra y avoir des orogénies simultanées aux deux extrémités d'un continent, l'une sera une vraie collision (avec chevauchements, obduction etc.) alors que l'autre sera plutôt un réajustement (ou "événement") tectonique en phase passive. L'inertie des masses continentales est telle que les plaques ne changeront pas instantanément de direction de mouvement ni de type de collision (pas de "flip-flop subduction" : BLAKE M.C.Jr., 1985). Il est capital de traîter ces différents régimes, s'ils existent, de manière différente.

. Les collisions ne surviennent pas toujours entre marges droites, avançant en directions exactement opposées. Les bordures sont souvent arquées ou irrégulières, comportant des saillants et/ou des rentrants ; elles peuvent se heurter obliquement et peuvent également intégrer des blocs continentaux précédemment détachés (des "soldats perdus"). L'on verra donc au sein d'un orogène des effets compressifs actifs et passifs juxtaposés, sans faire appel à des changements de régime. Ceci peut causer des mouvements de cisaillement non seulement en marge mais aussi à l'intérieur des continents durant et après leur collision et la fermeture des aires océaniques intermédiaires. (Voir à ce sujet les théories de PAVONI-1961 et de TAPPONIER et MOLNAR-1975).

. Les zones de faiblesse intra-cratoniques réagiront de façon différente aux régimes tectoniques successifs et par conséquent elles changeront le sens de leur mouvement et leur rejet, ainsi que l'intensité de leur activité à travers les temps. Mais elles resteront présentes et influenceront, quoique d'une façon de plus en plus secrète, les événements structuraux superficiels ;

"Old faults never die" (W.W. MALLORY-1982).

. L'interaction mécanique des étirements et coulissages intracontinentaux et des lignes structurales héritées expliquent de façon satisfaisante - et simple - l'ensemble des phénomènes décrits jusqu'ici, et permet d'intégrer certains phénomènes plus vastes, comme la constance de certains traits géologiques tels les dépocentres, ou certains changements de style à l'intérieur des zones orogéniques. Il n'y a pas besoin de faire appel comme cause initiale à des régimes thermiques ; ceux-ci sont réels mais conséquents.

. Les continents, en dérivant, peuvent incorporer des blocs océaniques et des prismes sédimentaires le long de leur marge active, aussi bien que perdre des blocs le long de leur marges passives, dont certains seront récupérés lorsque la marge devient active : l'on voit donc un cycle d'écartèlement et d'agglutination de blocs marginaux. La dislocation des marges passives suit le découpage originel des continents. Le grain structural des pièces continentales détachées influencera ainsi la paléogéographie des eugeoclinaux qui les accueillent ; elles pourront réapparaître en forme de massifs intramontagneux (haut-fonds, rides) lors de la compression orogénique suivante. Les petits mouvements en rotation ainsi que les jeux latéraux de tels blocs peuvent causer des plis précoces à l'intérieur des complexes orogéniques ; les anciens systèmes cisaillants pourront ainsi regagner une expression, atténuée mais réelle, dans les orogènes récents.

IX-A) FIGURES

X. RECOMMANDATIONS DE TRAVAIL

- Travaillez toujours en TROIS DIMENSIONS - faites des bloc-diagrammes plutôt que des coupes et/ou des cartes seules.

- N'oubliez pas la QUATRIEME DIMENSION : LE TEMPS. Tâchez de déchiffrer le déroulement des événements successifs, ne vous contentez pas d'imputer toutes les déformations à une seule phase tectonique (la dernière). Essayez de distinguer entre l'effet de l'HERITAGE et les apports des périodes structurales successives.

- Pensez REGIONAL. Evitez de faire de la géologie style timbre-poste. Plus d'un problème local "insoluble" trouve sa solution lorsqu'il est replacé dans son contexte régional : d'un ordre de grandeur supérieur.
  - N'hésitez pas de faire des comparaisons avec les bassins du même type le plus proches même si "plus proche" signifie en fait "assez lointain".

- Servez-vous des résultats de TOUTES LES METHODES accessibles, qu'elles soient anciennes ou récentes, chères ou bon marché, telles les méthodes potentielles (gravimétrie, magnétisme), la sismique réfraction, la morphologie et n'oubliez pas le fondement de la géologie ; l'observation de terrain.

- "SUIVEZ LE SOCLE". Gardez un oeil sur les niveaux profonds, traquez-les sur tous vos documents sismiques, même si cela vous demande de réinterpréter vos études ou retraiter vos profils.

- Appliquer des modèles est bon, parfois nécessaire mais ne remplace pas l'ANALYSE et la REFLEXION pragmatique. Ne vous laissez pas emporter vers des traitements sophistiqués, malgré l'attrait intellectuel qu'ils présentent, avant d'avoir établi votre base d'information aussi large que possible, bien que ceci soit un travail rébarbatif.

- Tenez compte de toute l'information existante, mais PENSEZ INDEPENDAMMENT. "Si vous voulez remonter à la source, vous devez avancer à contre-courant" (d'après VOLTAIRE).

XI. REMERCIEMENTS

Je remercie la DIRECTION EXPLORATION de la S.N.E.A.(P.) pour le support matériel accordé pour la préparation de ce Séminaire.

Je désire exprimer ma reconnaissance à tous mes anciens collègues et amis de longue date dans le Groupe ELF-Aquitaine et ses filiales qui, avec leurs conseils et discussions -ou tout simplement avec leur patiente attention- m'ont beaucoup aidé à élaborer ma philosophie et à réaliser ce rapport, en particulier à J.P. RICHERT et aux membres du Groupe de Géologie Structurale. J. GIRAUDMAILLET, P. VERNHES J. HENRY et R. RICATEAU m'ont appuyé lors de mes premières tentatives dans ce domaine, tandis que Sylvette TRIBOULET, J.R. STEVAUX, J. de la SERVE et G. NELY ont contribué à l'amélioration des versions successives du texte français.

Au cours de mes voyages et séjours en Amérique du Nord, J. HAUN, W.W. MALLORY, R. de VOTO, Robbie GRIESS et Annie COURY (tous de Denver), ainsi que P. HOFFMAN (Ottawa) ont partagé avec moi leur temps, leurs idées et leurs documents. J. WALPER (Ft. Worth), N. PAVONI (Zurich), D. RIGASSI (Genève) et P.V. ZALAN (Rio de Janeiro) m'ont également fourni des données inédites.

J. FANSHAWE (Billings) m'a introduit à la tectonique de l'avant-pays des Rocheuses et m'a permis de profiter de sa grande connaissance et de sa sagesse géologiques.

J. SCHOEFFLER (Pau), N.J. Mc MILAN (Calgary), J. DERCOURT (U. de Paris VI), M. MATTAUER (U. de Montpellier), R. BOURROUILH (U. de Bordeaux) et D. RIGASSI (Genève) m'ont donné leur soutien à des moments critiques de l'évolution de mes travaux.

Je désire associer à cet ouvrage ANDRE ZOLNAI (géologue à Calgary) qui fut mon principal conseiller, critique, et mon correcteur pour les deux textes ; il m'a également fourni beaucoup d'idées et d'observations de terrain et il m'a procuré bon nombre de documents.

La responsabilité des vues (et hérésies ?) contenues dans les Notes ci-dessus reste cependant entièrement la mienne.

XII. LISTE BIBLIOGRAPHIQUE

BAARS D.L. & STEVENSON G.M. (1982)
Subtle Stratigraphic traps in Paleozoic rocks of Paradox Basin. in : AAPG. Mem. 32. pp. 131-158

BALDUCCI A. & POMMIER G. (1970)
Cambrian oil field of Hassi Messaoud, Algeria in A.A.P.G. Memoir n° 14, pp. 477-488

BARTLETT W.L., FRIEDMAN N. & LOGAN J.M (1981)
Experimental folding and faulting of rocks under confining pressure Tectonophysics, 79, pp. 255-277

BERGERAT F. (1977)
La fracturation de l'avant-pays jurassien entre les fossés de la Saône et du Rhin. Analyse et essai d'interprétation dynamique.
Rev. Géog. Phys. Géol. Dyn. Vol. XIX, fasc. 4, pp. 325-338

BOURROUILH R., RICHERT J.P. et ZOLNAI G. (1988)
The North Pyrenean Intraplate Basin : evolution and hydrocarbons, and Iberia versus Europe : effects of Continental Break-up and Round-up on hydrocarbon habitat.
A.A.P.G. Bull, V. 72, n° 8, p. 990 (Résumés seulement)

BROWN W.G. (1984)
Basement involved tectonics, foreland areas
A.A.P.G. Continuing education course note series n° 26

BRUN J.P. & CHOUKROUNE P. (1983)
Normal faulting, block tilting and décollement in a stretched crust.
Tectonics V.2 n° 4 pp. 345-356

CLEMENT J.H. (1987)
Cedar Creek : A significant Paleotectonic feature of the Williston Basin
A.A.P.G. Memoir n° 41, (J.A. PETERSON Ed.), pp. 213-240

COOK D. (1983)
The Northem Franklin Mountains, N.W.T., Canada - a scale-model of the Wyoming Province.
Rocky Mtn. Assoc. Géol. :"Rocky mountain Foreland Basins and Uplifts" pp. 315-338

CRAWFORD F.D., SZELEWSKI C.E. & ALVEY G.D. (1985)
Geology and exploration in the Takutu graben of Guyana and Brazil.
Journ. of Petr. Geol. V. 8 n° 1, pp. 5-36

CROWELL J.C. (1962)
Displacement along the San Andreas Fault, California in : "Special G.S.A. Papers" n° 71

CROWELL J.D (1974)
Origin of Late Cenozoic basins in Southern California in : "Tectonics & sedimentation" (W.R. Dickinson ed.).
S.E.P.M. Spec. Publ. 22, Tulsa. pp. 190-204

DE VOTO R.H. (1980)
Pennsylvanian stratigraphy and history of Colorado - Rocky Mtn. Assoc. Geol.
Symposium pp. 71-101

FINLAYSON D.M., LEVEN J.H. & ETHRIDGE M.A. (1988)
Structural styles and bassin evolution in Eromanga Region, Eastern Australia.
A.A.P.G. Bull V. 72, n°1, pp. 33-48

FOX F.G (1985)
Structural geology of the Parry Island fold belt (Arctic Canada).
Bull. Can. Petr. Geol. V. 33, n° 3, pp. 306-340

GATEWOOD L.E. (1970)
Oklahoma City field - anatomy of a giant. in : AAPG. Memoir 14 "Geology of giant petroleum fields", (M.E. Halbouty ed.)
pp. 223-254

GELARD J.P. & RAT P.(1977)
Vues actuelles sur les grandes lignes de la structure de la Bourgogne en relation avec l'évolution post-hercynienne.
Vème réunion annuelle des Sc. de la Terre - Rennes - Abstracts p. 243

GELARD J.P. (1978)
La fracturation de la Bourgogne méridionale, essai d'interprétation et implications tectoniques régionales.
Rev. Géogr. Phys. Géol. Dyn. (2) Vol. XX. Fasc.1. pp. 5-28

GWYNN T.A. (1964)
The Cedar Creek Anticline.
3rd internatl. Williston Basin Symposium pp. 192-199

HAM W.E., DENISON R.D. and MERRITT C.A. (1964)
Basement rocks and structural evolution - Southern Oklahoma
Okla. Geol. Surv. Bull. 95

HARDING T.P. (1974)
Petroleum traps associated with Wrench Faults.
A.A.P.G. Bull V. 58 n° 7, pp 1290-1304

HARDING T.P. & LOWELL J.D. (1979)
Structural styles, their plate tectonic habitats and hydrocarbon traps in petroleum provinces.
A.A.P.G. Bull V. 63, n° 7, pp 1016-1058

HARDING T.P. (1983)
Divergent wrench fault and negative flower structure, Andaman sea. in : (A.W. Bally ed.) "Seismic expression of structural styles".
A.A.P.G. Studies in Geology Series N°15. pp. 4.2-1 to 4.2-8

HARDING T.P., GREGORY R.F. & STEPHENS L.M. (1983)
Convergent wrench fault and positive flower structure, Ardmore Basin, Oklahoma.
Ibid. pp. 4.2-13 to 4.2-17

HENRY J. (1983)
Manuel d'analyse structurale (Dans : Méthodes modernes de géologie de terrain, n° 2a). Editions TECHNIP.

HOFFMAN P.F., BELL I.R., HILDEBRAND R.S. and THORSTAD L. (1977)
Geology of the Athapuscow Aulacogen, East Arm of Great Slave Lake, District of Mackenzie.
Geol. Surv. of Canada, Rep. of A ct., 77-1/A, pp. 117-129

JOHNSON W.D. & SMITH H.R. (1964)
Geology of the Winnett - Mosby area... Montana
U.S.G.S. Bulletin n° 1149

KINGMA J.T. (1958)
Possible origin of piercment structures, local unconformities and secondary basins in the Eastern Geosyncline, New Zealand
N.Z. Journ. of Geol. & Geoph. V.1, n° 2, pp. 269-274

LAROCHE P.J. (1983)
Appalachians of Southern Quebec seen through seismic line n° 2001 - in (A. W. Bally ed.) "Seismic expression of structural styles". A.A.P.G Studies in Geology Series n° 15, pp. 3.2.1-7 to 3.2.1-22

LOWELL J.D. (1972)
Spitzbergen Tertiary Orogenic Belt and the Spitzbergen Fracture Zone. Geol. Soc. Am. Bull V. 83, n° 10, pp. 3091-3102

MATTAUER M. (1980)
Les déformations des matériaux de l'écorce terrestre
Hermann Paris (ed.) - Collection Méthodes, 2è édition

MATTE Ph. (1986)
La chaîne Varisque... au Permo-Carbonifère
Bull. Soc. Géol. Fr. (8) t. II n° 1, pp. 9-24

MEILLIEZ F. (1987)
Géologie du Massif Armoricain
Géochronique n° 21, p. 10

MOLNAR P. & TAPPONIER P. (1975)
Cenozoic tectonics of Asia : effects of a continental collision.
Science V. 189, n° 4201, pp. 419-426

MOODY J.D. & HILL M.J (1956)
Wrench-fault tectonics.
Geol. Soc. Am. Bull. V. 67, pp. 1207-1246

MOODY J.D. (1973)
Petroleum exploration aspect of wrench-fault tectonics.
A.A.P.G. Bull., V. 57, n° 3, pp. 449-476

MUEHLBERGER W.R. (1980)
The Shape of North America during the Precambrian, in : "Continental Tectonics".
Nat'l Acad. Sc. Geoph. Res. Board., p. 175-182

OKULITCH A.V., PACKARD J.J. & ZOLNAI A.I. (1986)
Evolution of the Boothia Uplift, arctic Canada.
Can. Journ. Earth Sci. V. 23, n° 3, pp. 350-358

PATTERSON J.G. (1986)
The Amer Belt ; remnant of an Aphebian foreland fold and thrust belt.
Can. J. Earth Sci. V. 23, pp. 2012-2023

PAVONI N. (1961)
Faltung durch Horizontalverschiebung.
Ecl. Geol. Helv. V. 54, n° 2, pp. 515-534

PRITCHETT R.W. (1985)
Seismic Profiles of the Western Laramie Basin-Wyoming in : "Seismic Exploration of the Rocky Mountain Region" (R.R. Gries & R.C. Dyer editors) R.M. Assoc. Geol. & Denver Geoph. Soc. pp. 225-232

RAY R.R. & KEEFER W.R (1985)
Wind River Basin, Central Wyoming - in : "Seismic Exploration of the Rocky Mountain Region" (R.R. Gries & R.C. Dyer editors)
R.M. Assoc. Geol. & Denver Geophys. Soc. pp. 201-212

READING H.G. (1980)
Characteristics and recognition of strike-slip fault systems.
Spec. Publ. Intern Assoc. Sedim. n° 4, pp. 7-26

RIEDEL W. (1929)
Zur Mechanik geologischer Brucherscheinungen
Zentralblatt Min., Geol., Pal. - Abt.B.

RIGASSI D.A. (1986)
Wrench-faults as a factor controlling petroleum occurences in West Siberia
A.A.P.G. Memoir n° 40,pp. 529-544

RODGERS D.A. (1980)
Analysis of pull-apart basin development produced by "en-echelon" strike-slip faults.
Spec. Publ. Int. Ass. Sedim. 4, pp. 27-41

ROLET J. (1984)
Graben losangiques (pull-apart) en régime de décrochement.
Le rôle des coulissements hercyniens dans l'individualisation des bassins carbonifères du Massif Armoricain.
Ann. Soc. Geol. Nord, Tome CIII, pp. 209-220

RUHLAND M. (1972)
Le rôle des décrochements dans le socle vosgien et en bordure du Fossé Rhénan. in : "Approaches to taphrogenesis" Procedings of Int. Rift Symp. Karlsruhe (13-15.IV.1972), pp. 167-171

SALES J.K. (1971)
Structure of the northern margin of the Green River Basin, Wyoming. Wyo. Geol. Assoc. Guidebook pp. 85-102

STEARNS D.W. (1971)
Mechanisms of drape folding in the Wyoming Province Wyo. Geol. Assoc. - 23rd Ann. Field Conf., pp. 125-143

STEARNS D.W. (1978)
A.A.P.G. Structure Seminar Notes (Nov. 22-23, Calgary), pp. 1-15

STONE D.S. (1974)
Lineaments, their role in tectonics of Central Rocky Mountains. Wyo. Geol. Assoc. Earth Sc. Bull. V. 7, n° 4, pp. 1-11

SYLVESTER A.G. & SMITH R.R. (1976)
Tectonic transpression and basement-controlled de formation in San Andreas fault. zone, Salton trough, California. A.A.P.G. Bull. V. 60, n° 12, pp.

TAPPONIER P. PELTZER G. & ARMIJO R. (1986)
On the mechanics of the collision between India and Asia-In : Collision Tectonics (Coward M.P. & Ries A.C. eds). Geol.Soc.Spec. Publ. n° 19, pp. 115-157

TCHALENKO J.S. (1970)
Similarities between Shear Zones of different magnitudes. Geol.Soc.Am. Bull. V. 81, pp. 1625-1640

THOMAS G.E. (1971)
Continental plate tectonics ; Southwest Wyoming. Wyo.Geol.Assoc. Guidebook pp. 103

THOMAS G.E. (1974)
Lineament-block block tectonics ; Williston - Blood Creek Basin. A.A.P.G. Bull. V. 58, n° 7, pp. 1305-1322

TUZO WILSON J. (1965)
A new class of Faults and their bearing ou Continental Drift.
Nature n° 4995, pp. 343-347

U.S.G.S (1975)
Paleotectonic investigations of the Pennsylvanian system in the U.S.
Professional Paper 853 (Mc Kee & Crosby ed.)

WALPER J.L. (1977)
Paleozoic tectonics of the Southern margin of North America
Gulf Coast Assoc. Geol. Soc. 27th ann-mtg
Austin (TEx) 26-28-X-77, pp. 230-241

WEGMAN E. (1961)
Anatomie comparée des hypothèses sur les plissements de couverture (le Jura plissé).
Bull. Géol. Inst. Univ. Uppsala, n° 40, pp. 169-182

WILCOX R.E., HARDING T.P. & SEELY D.R. (1973)
Basic Wrench Tectonics.
A.A.P.G. Bull V. 57, n° 1, pp. 79-96

WISE D.U. (1963)
An outrageous hypothesis for the tectonic pattern of the North American Cordillera.
Geol. Soc. Am. Bull. V. pp. 357-362

ZALAN P.V. (1987)
Identification of Strike-Slip faults in seismic sections.
Soc. Expl.Geoph. 57 th Ann. Mtg.
New Orleans 1987 pp. 116-118

ZOLNAI G. (1971)
Le front Nord des Pyrénées Occidentales
Histoire Structurale du Golfe de Gascogne - Technip Ed., pp. IV-5, 1-9

ZOLNAI G. (1975)
Sur l'existence d'un réseau de failles de décrochement dans l'avant-pays Nord des Pyrénées Occidentales (Wrench-faults in the northern foreland of the western Pyrénées-France).
Rev. Géogr. Phys. Géol. Dyn. Vol. XVII. fasc. 3, pp. 219-238

ZOLNAI G. (1982)
Late Mississippian-Pennsylvanian Orogenic Movements of West Canadian Platform and Adjacent Areas - Their role in sedimentation and hydrocarbon accumulation.
A.A.P.G. Bull., V. 66, n° 5, pp. 646-647 (Résumés seulement).

ZOLNAI G. (1985)
Les aulacogènes du continent Nord-Américain.
Bull. Soc. Géol. Fr. V.8, t. II, n° 5, pp. 809-818

ZOLNAI G. (1989)
Arches, Lineaments, mega-shears, rifts and aulacogens of North America, to be published in the : Bull des Centres de Recherche - ELF Aquitaine, in May 1989.

FIGURES

I. BASICS

MECANISMES DE BASE

I-B) FIGURES

		Fig. n°
Definition :		
. Dip-slip, Strike-slip, Oblique-slip	/MATTAUER/	I-1
. Dextral-senestral	/G.Z./	I-2
. Clockwise-anticlockwise	/G.Z./	I-3
. Rotation - tilting	/G.Z./	I-4
Fault systems :		
. "Basic" fracture-pattern	/RIEDEL, TCHALENKO/	I-5
. Shear ellipsoid (faults & folds)	/HARDING/	I-6
. Rhomb-diagram	/CROWELL/	I-7
. Shear zones	/WILCOX, KINGMA/	I-8
. Off-sets (right-stepping left-stepping)	/CROWELL/ /RODGERS/	I-9
. Transcurrent faults	/CROWELL/	I-10
. Transform faults	/TUZO WILSON/	I-11
. Transpression - Transtension	/SYLVESTER/	I-12/A
Flower-structures - positive & negative	/LOWELL/ /HARDING/	I-12/B-C
. Open (extensional) fractures	/MATTAUER, HENRY/	I-13
. Pure shear vs. simple shear	/THOMAS/	I-14
What is not a wrench-fault ?	/G.Z./	I-15
Associated folds :		
. Drape folds	/G.Z./	I-16
. Drag-folds	/G.Z./	I-17
. Free folds - Forced folds	/G.Z./	I-18

I-B) FIGURES

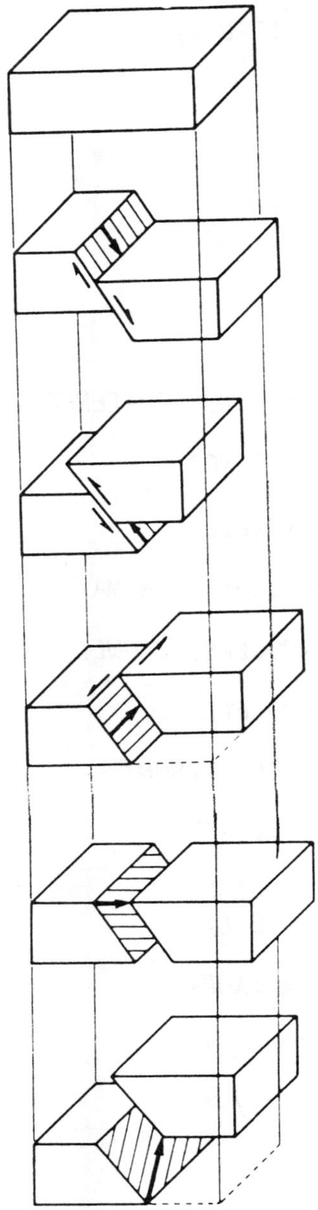

Bloc-diagramme des différents types de failles.
1) faille normale ; 2) faille inverse ; 3) décrochement senestre ; 4) décrochement senestre normal ; 5) décrochement senestre inverse.

Block-diagram showing the various types of faults :

1) normal fault
2) Reverse fault
"dip-slip" faults

3) left-lateral wrench-fault - "strike-slip" fault

4) left-lateral "normal wrechch"
5) left-lateral "reverse wrench"
"oblique-slip" faults

From/d'après : M. MATTAUER (1980)

Fig. I-1 - Dip-slip Strike-slip Oblique-slip (MATTAUER)

Fig. I-1 - Failles à rejeu vertical, oblique et horizontal (MATTAUER)

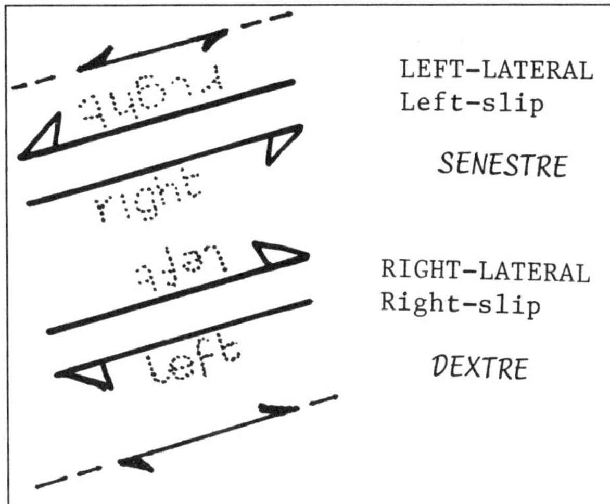

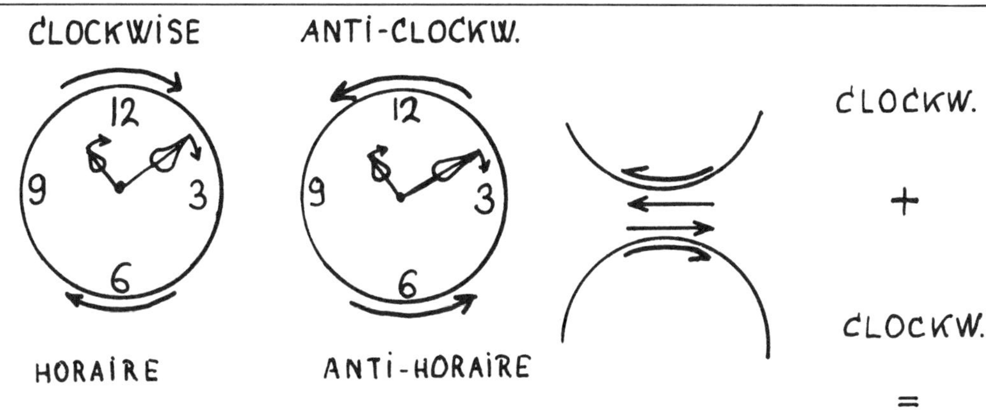

Horaire - Anti-horaire

Fig. I-3 Clockwise - Anticlockwise

When both of the opposite trains go "to the right", the relative displacement between them is left-lateral.

It remains left-lateral between the two trains going in the same direction, since one of them is progressing faster than the other.

Lorsque chacun des deux trains opposés va "à droit", le mouvement relatif est senestre.

Le déplacement relatif reste senestre également entre les deux trains allant dans la même direction, puisque l'un avance plus vite que l'autre.

Fig. I-2 Dextral - Senestral

Dextre - Senestre

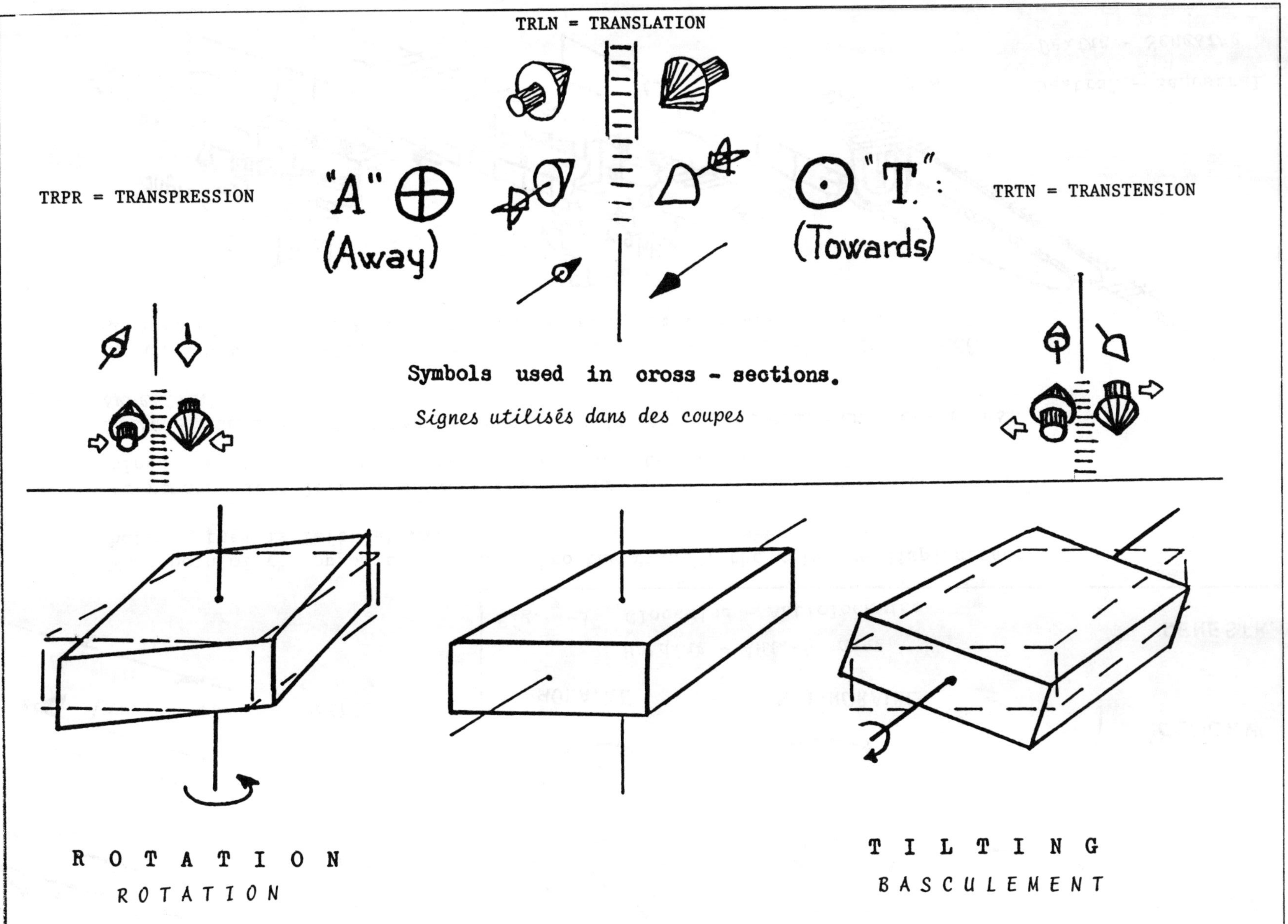

Fig. I-4 - Rotation - tilting

Fig. I-4 - Rotation - Basculement

I-5/C

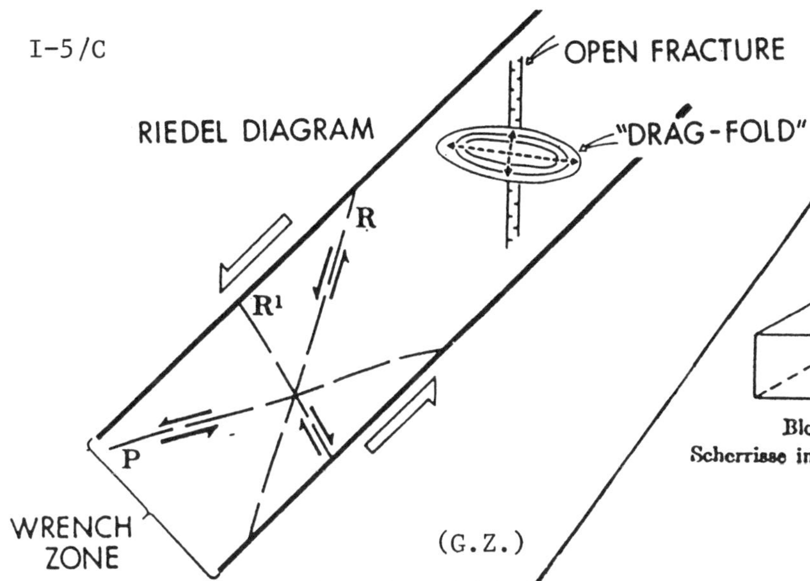

(G.Z.)

I-5A RIEDEL (1929)

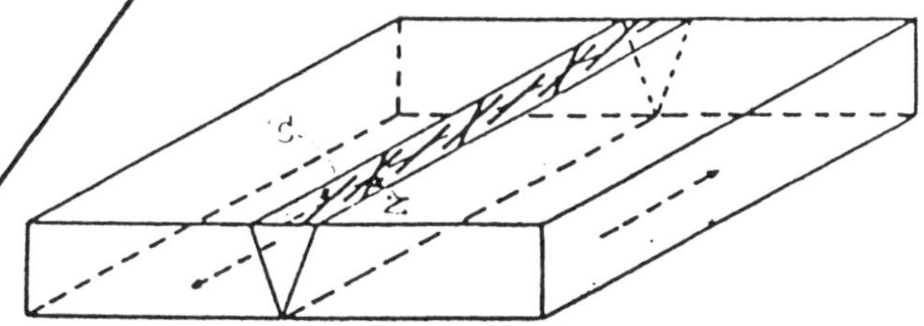

Blockdiagramm zur Veranschaulichung der Lage der Zug- und Scherrisse in den Versuchen. Z: Zugrisse, S: Scherrisse, Keilförmige Störungszone.

flachere steilere
Staffelung von Flächen.

Z : Zugrisse = drag-fractures = fractures d'entraînement

S : Scherrisse = shear-fractures = fractures de cisaillement

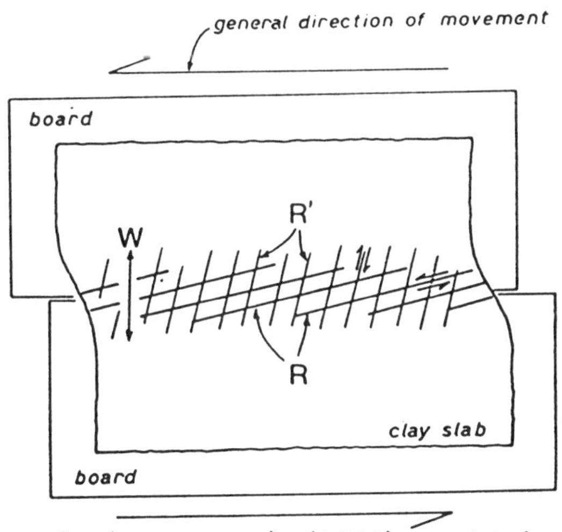

I-5/B (TCHALENKO) (1970) (plan)

Fig. I-5 - "Basic" fracture-pattern (RIEDEL, TCHALENKO)

Fig. I-5 - Jeu de fractures (RIEDEL, TCHALENKO)

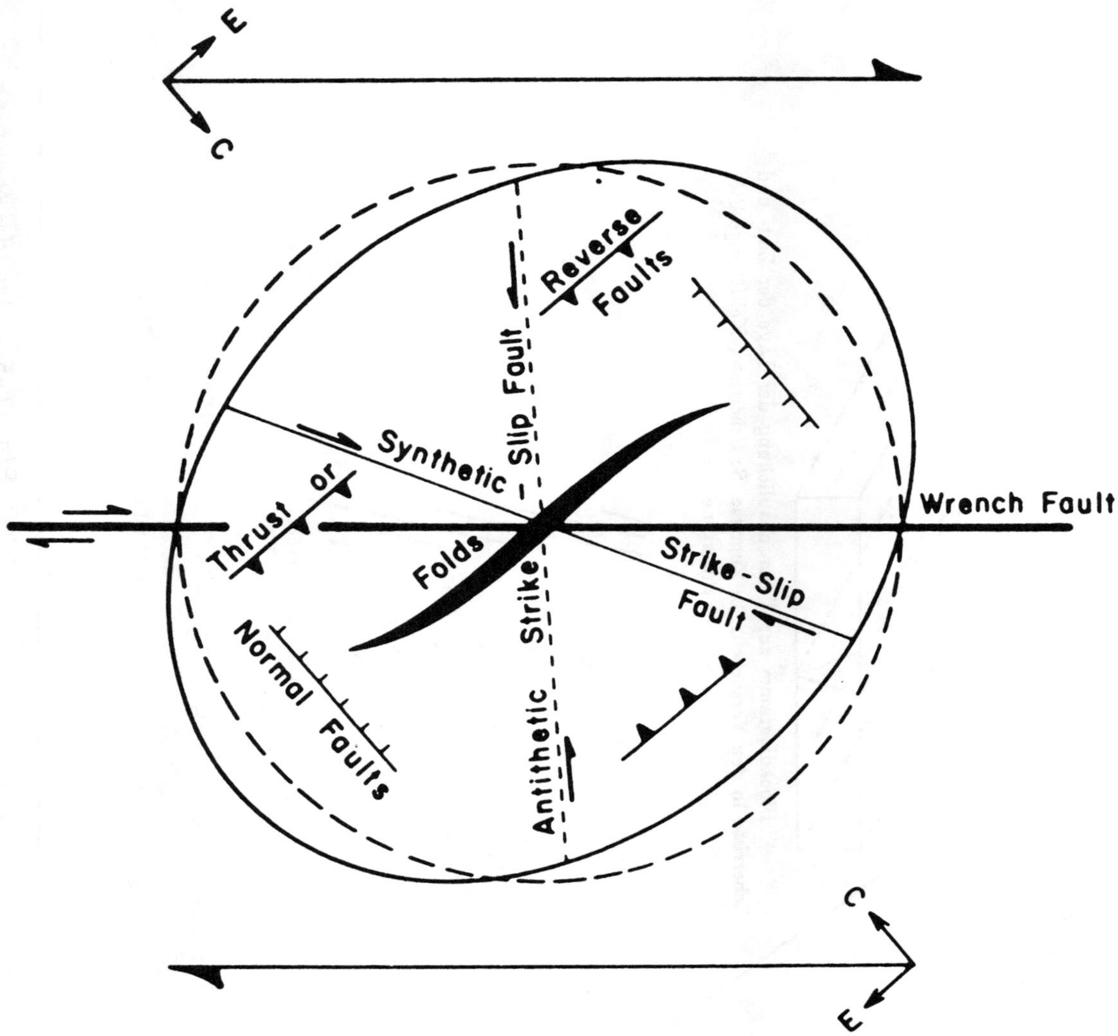

—Geometric relation of folds and faults to right-slip wrench fault combined schematically with strain ellipse and principal strain directions, contraction, *C*, and extension, *E*. Axial traces of folds have flattened S-shapes, corresponding to field and laboratory observations. Geometry for left-slip wrench fault is mirror image of this diagram. Adapted from Harding (1974, p. 1291) - In: SYLVERSTER et al.-1976

Fig. I-6 - Shear ellipsoid (faults & folds)
(HARDING)

Fig. I-6 - Ellipsoïde de cisaillement
(HARDING)

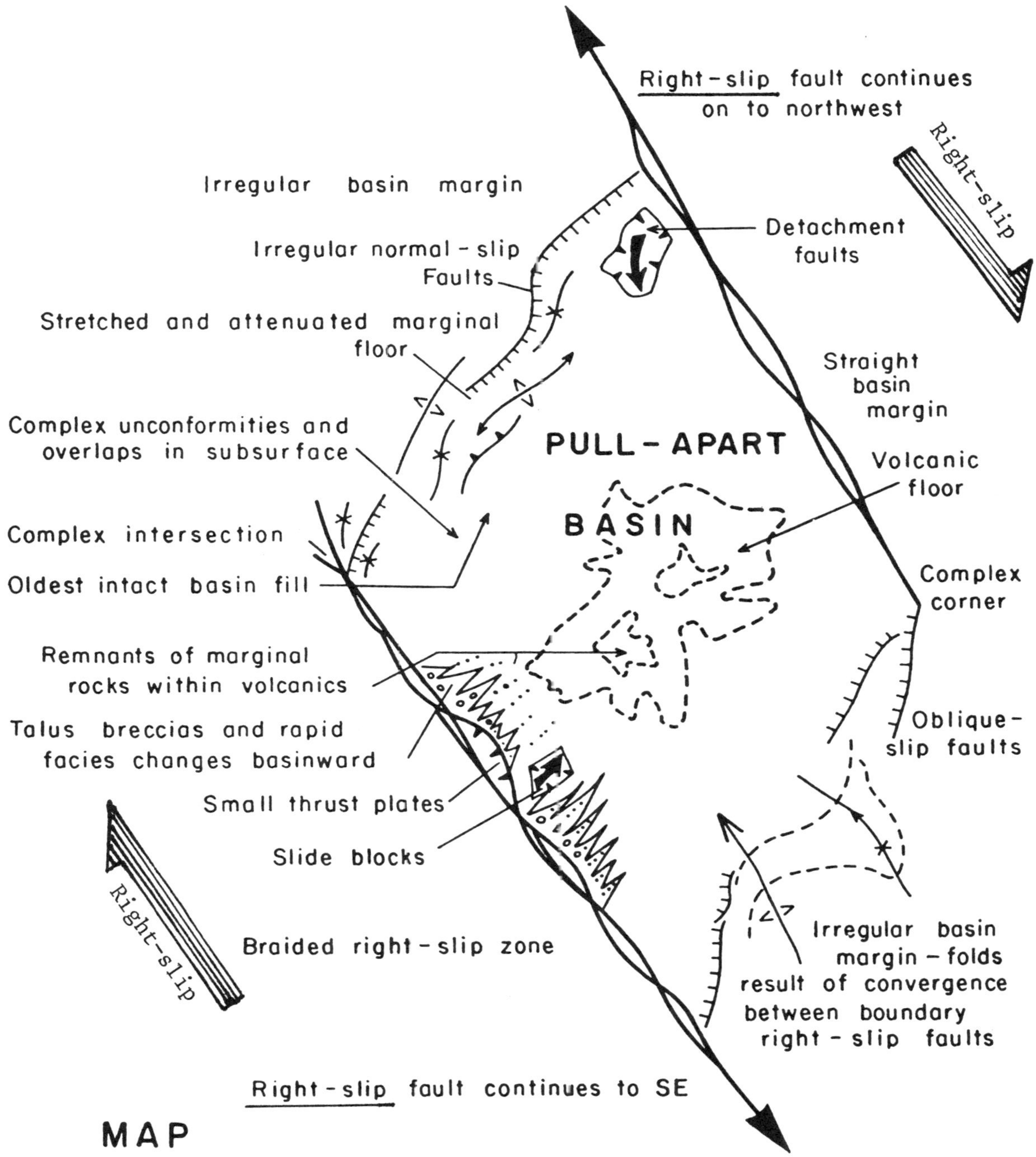

Sketch map of idealized pull-apart basin

CROWELL (1974)
(completed/complété)

Fig. I-7 - Rhomb-diagram
(CROWELL)

Fig. I-7 - Diagramme rhombique
(CROWELL)

Ronald E. Wilcox, T. P. Harding, and D. R. Seely (1973)

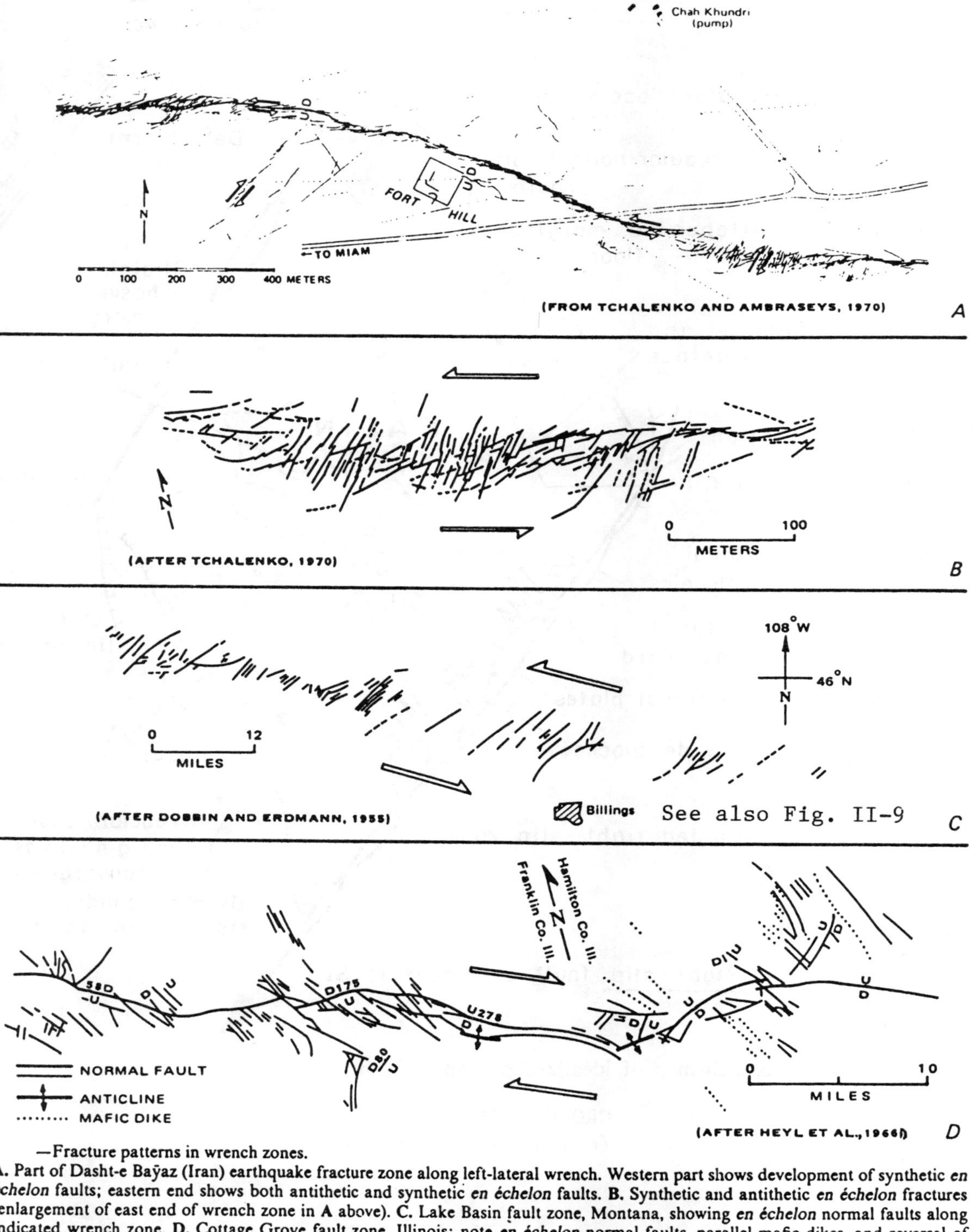

—Fracture patterns in wrench zones.
A. Part of Dasht-e Bayaz (Iran) earthquake fracture zone along left-lateral wrench. Western part shows development of synthetic *en échelon* faults; eastern end shows both antithetic and synthetic *en échelon* faults. **B.** Synthetic and antithetic *en échelon* fractures (enlargement of east end of wrench zone in **A** above). **C.** Lake Basin fault zone, Montana, showing *en échelon* normal faults along indicated wrench zone. **D.** Cottage Grove fault zone, Illinois; note *en échelon* normal faults, parallel mafic dikes, and reversal of vertical separation sense on throughgoing strike-slip fault. Vertical components indicated in feet.

Fig. I-8-A – Shear zones (WILCOX)

– Zones de cisaillement (WILCOX)

KINGMA (1958)

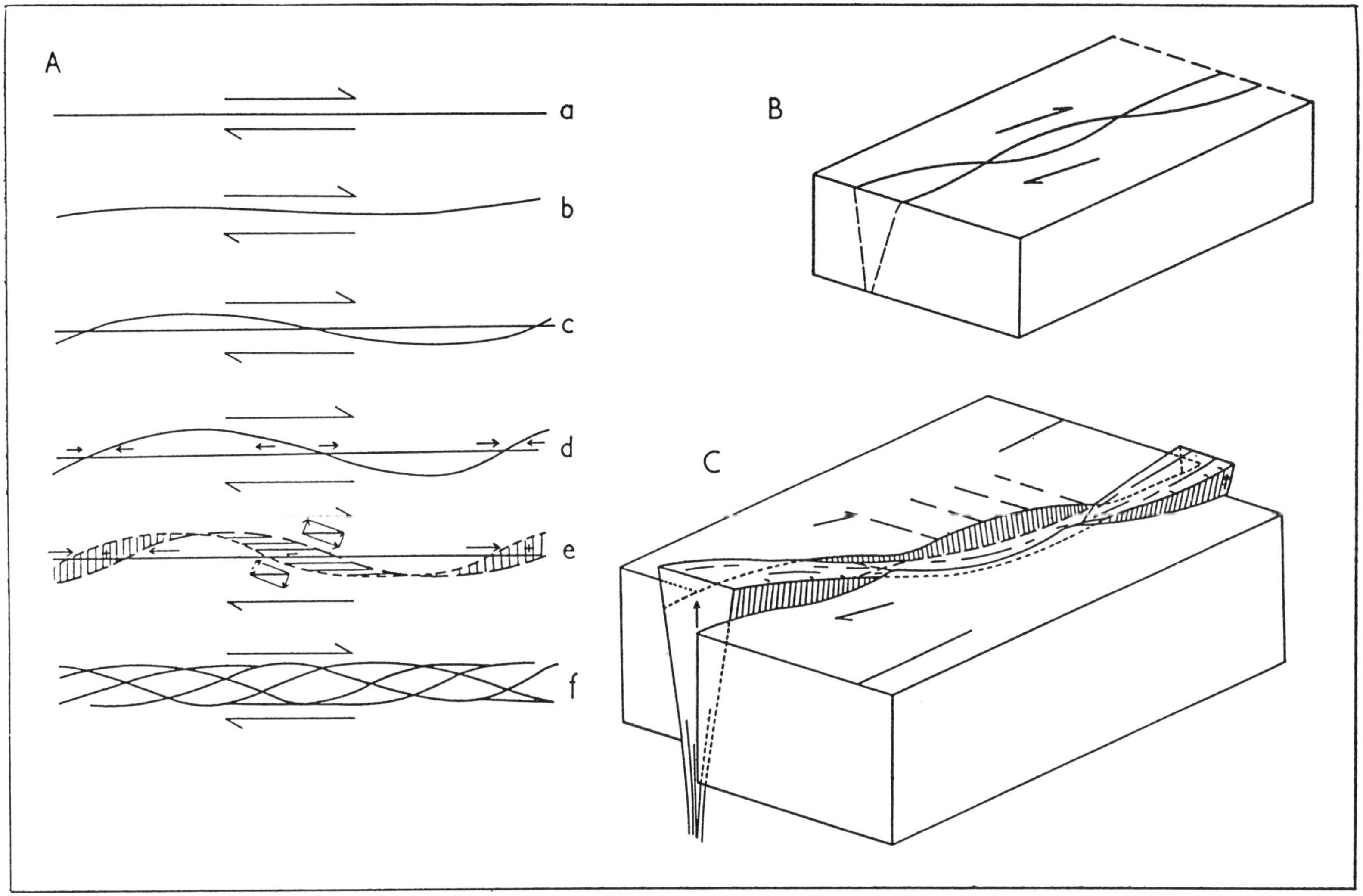

—A. Stages in deformation and in change of direction along part of a dextral transcurrent fault; — denotes tensional and sinking areas; + compressional and rising areas. B. Dextral transcurrent fault-zone before movement. C. same fault-zone after movement has taken place; development of horsts and graben.

Fig. I-8-B - Zones de cisaillement (KINGMA)

Fig. I-8-B - Shear zones (KINGMA)

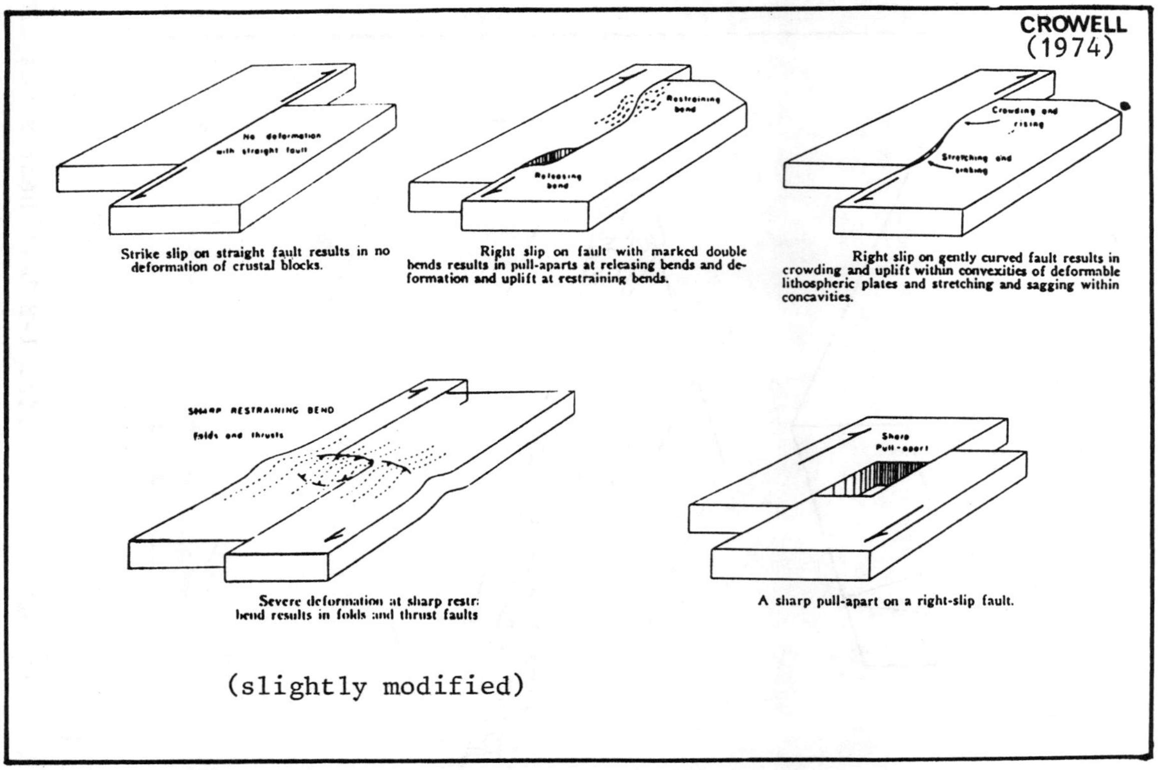

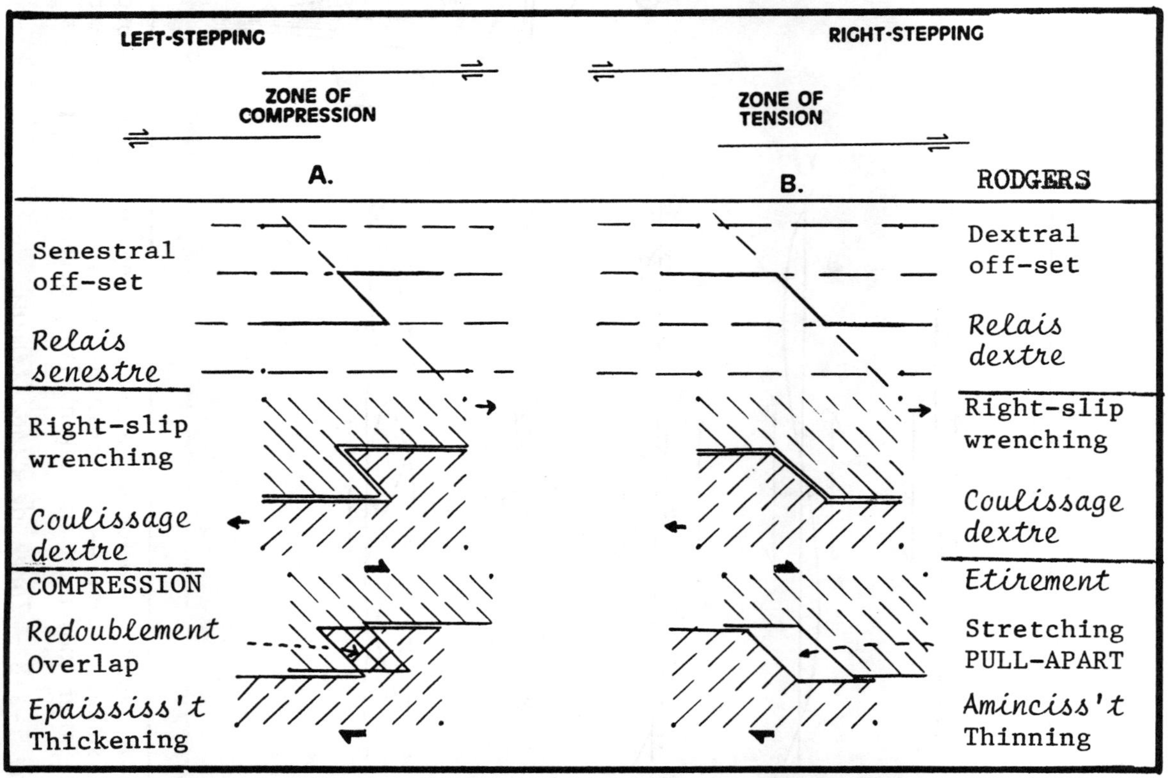

Fig. I-9 - Off-sets (right-stepping, left-stepping) (CROWELL, RODGERS)

Geological Society of America Special Paper 71

CROWELL, PLATE 1

(1962)

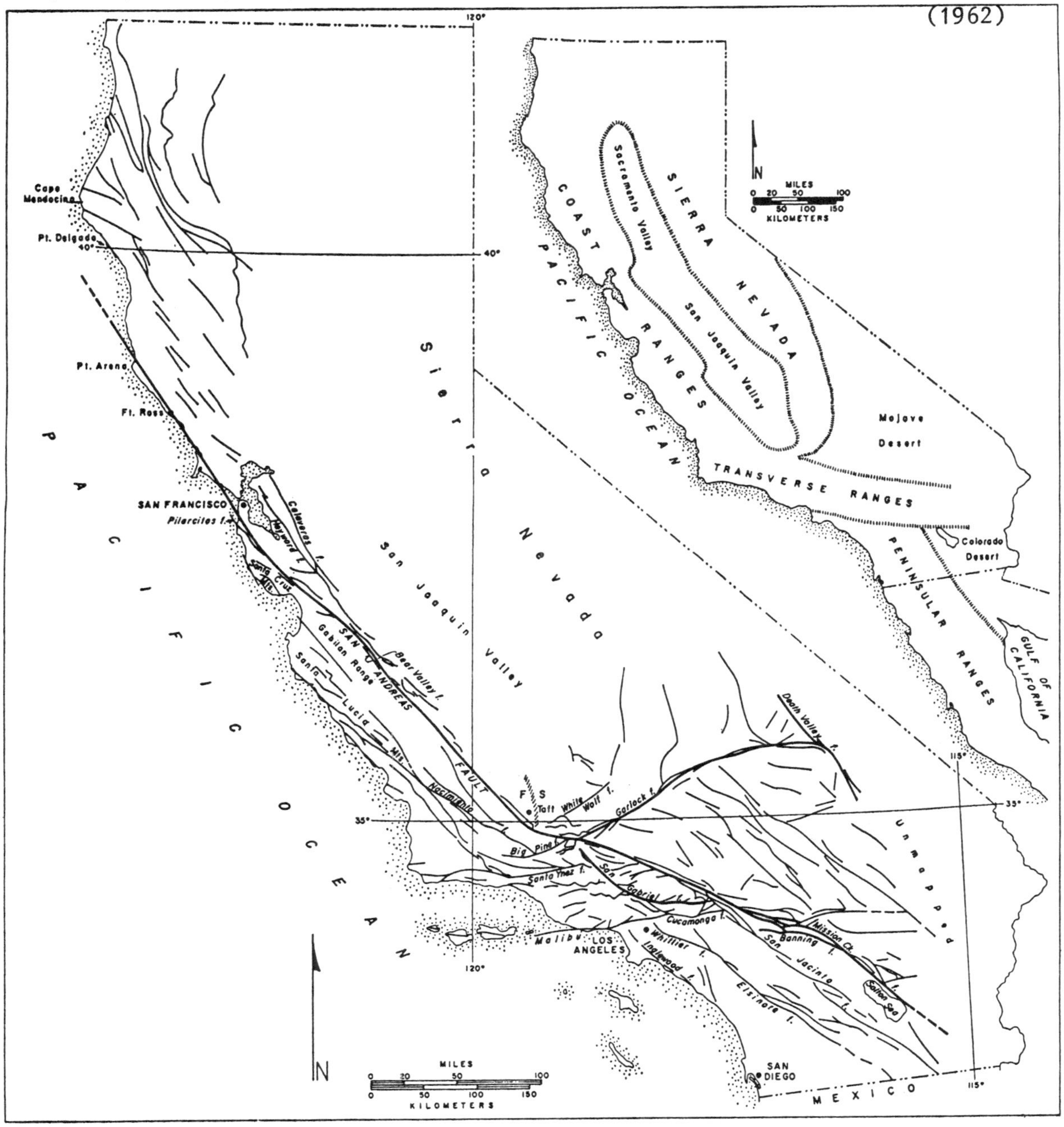

GENERALIZED FAULT MAP OF COASTAL AND SOUTHERN CALIFORNIA

Compiled and edited from selected data in Jenkins (1943), Jahns (1954), Irwin (1960), and Dibblee (1960a). Near Taft buried contact is between Franciscan (F) and Sierran (S) rocks. Principal physiographic provinces at upper right

Fig. I-10 - Transcurrent faults
(**Crowell**)

Fig. I-10 - Failles transcurrantes
(**Crowell**)

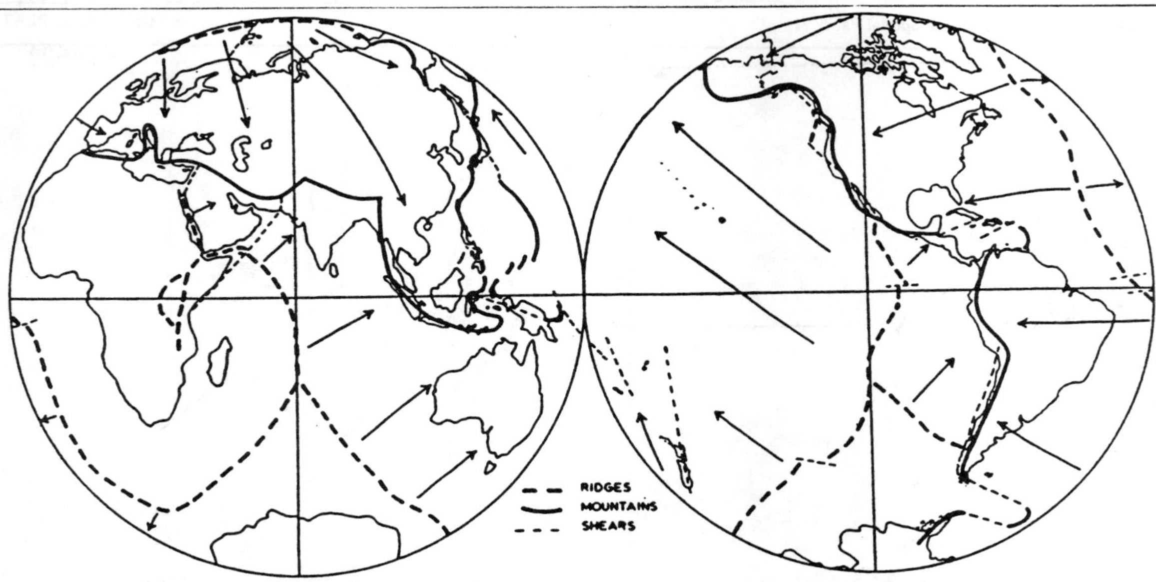

Sketch map illustrating the present network of mobile belts, comprising the active primary mountains and island arcs in compression (solid lines), active transform faults in horizontal shear (light dashed lines) and active mid-ocean ridges in tension (heavy dashed lines).

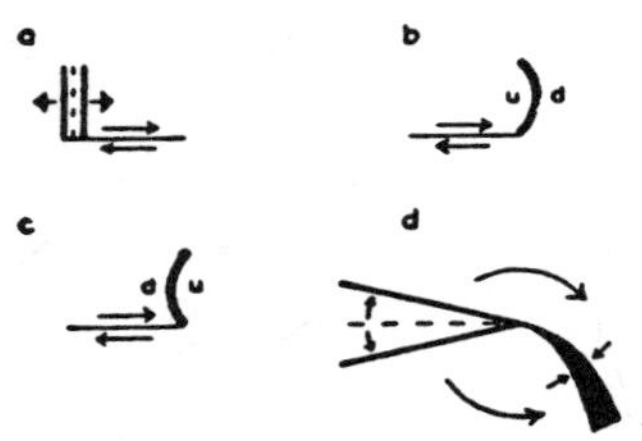

Diagram illustrating the four possible right-hand transforms, *a*, Ridge to dextral half-shear; *b*, dextral half-shear to concave arc; *c*, dextral half-shear to convex arc; *d*, ridge to right-hand arc.

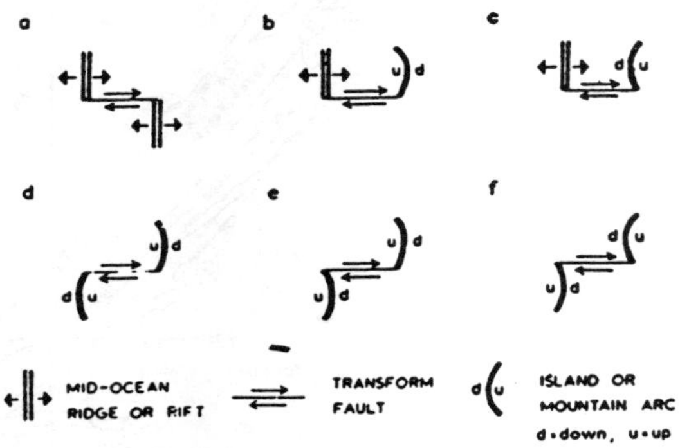

Diagram illustrating the six possible types of dextral transform faults. *a*, Ridge to ridge type; *b*, ridge to concave arc; c, ridge to convex arc; *d*, concave arc to concave arc; *e*, concave arc to convex arc; *f*, convex arc to convex arc. Note that the direction of motion in *a* is the reverse of that required to offset the ridge.

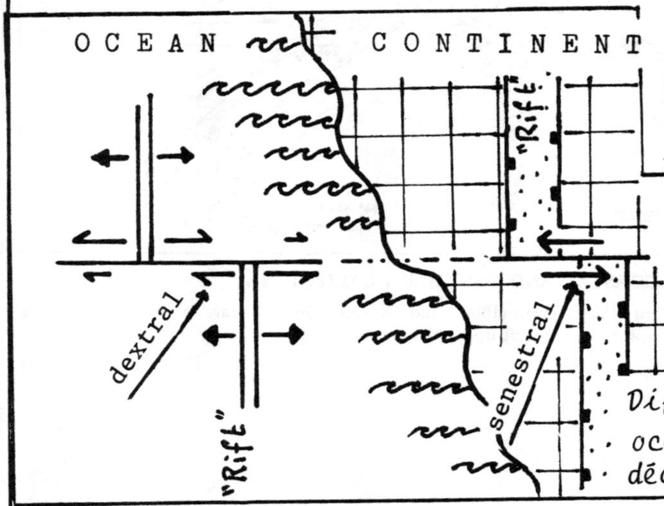

Difference between oceanic transform-faults and continental wrench-faults

Différence entre failles transformantes océaniques et failles de décrochement continentales

Fig. I-11 - Transform faults (TUZO WILSON)

Fig. I-11 - Failles transformantes (TUZO WILSON)

Arthur G. Sylvester and Robert R. Smith (1976)

- CALIFORNIA -

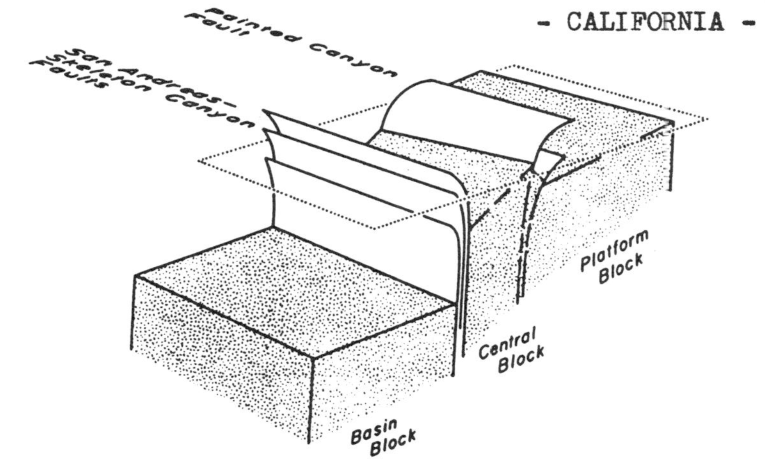

—Idealized block diagram of basement and principal faults in Mecca Hills. Dotted parallelogram represents surface.

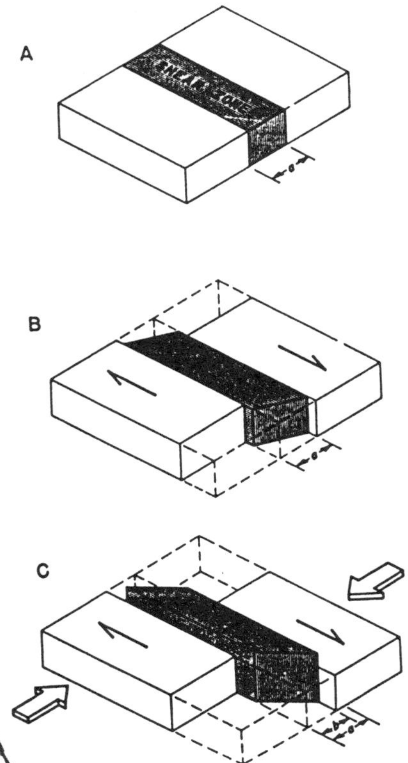

—Conceptual model of transpression basement deformation.

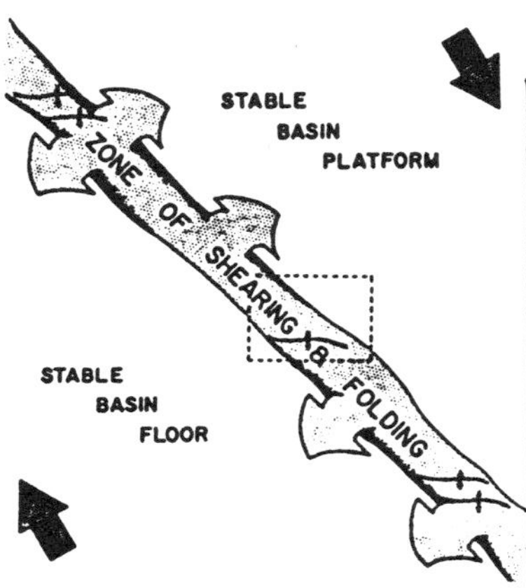

—Conceptual model of transpressive deformation, northeast margin of Salton trough. Rectangle shows study area relative to model.

LOWELL - 1972.
(Spitzbergen)

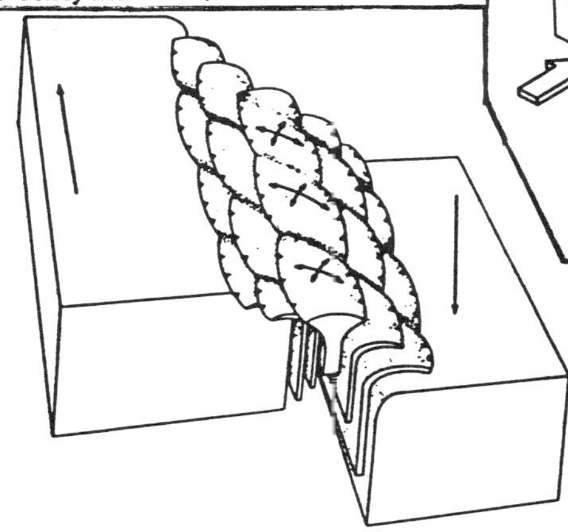

Figure 9. Conceptual diagram of an upthrust-bounded welt created by convergent strike-slip or transform motion. Portions of two plates or blocks are shown, the one moving at a low convergent angle into the other in the same plane, causing each segment to move a small distance successively past the other in a right-slip direction. A space problem is created, and the easiest direction of relief for the crowded material is upward. A welt with downward-tapering wedges and upthrust margins is created. The upthrusts are not necessarily so symmetrically disposed as shown and the deeper faults may tend to coalesce and braid rather than be parallel.

Folding precedes faulting in this type of deformation, which explains the observations by Orvin (1940) that the faults are younger than the folds in the Spitsbergen Tertiary orogenic belt.

Fig. I-12/A - Transpression - **Flower structures**
(SYLVESTER) (LOWELL)

Fig. I-12-A - Transpression - Structures en fleur
(SYLVESTER) (LOWELL)

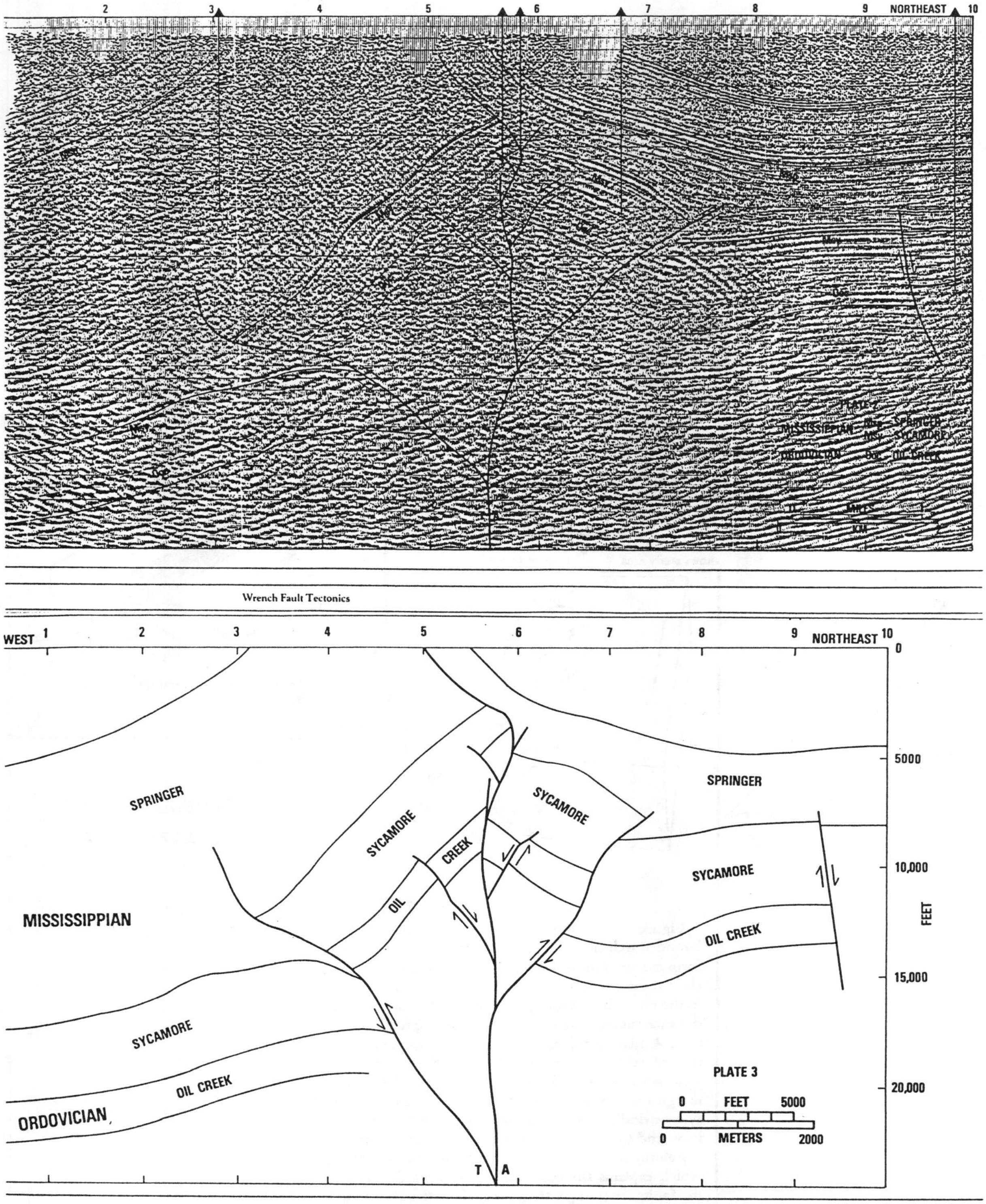

Fig. I-12/B - Flower-structures - Positive (HARDING, 1983) - *Structures en fleurs - Positives* (HARDING, 1983)

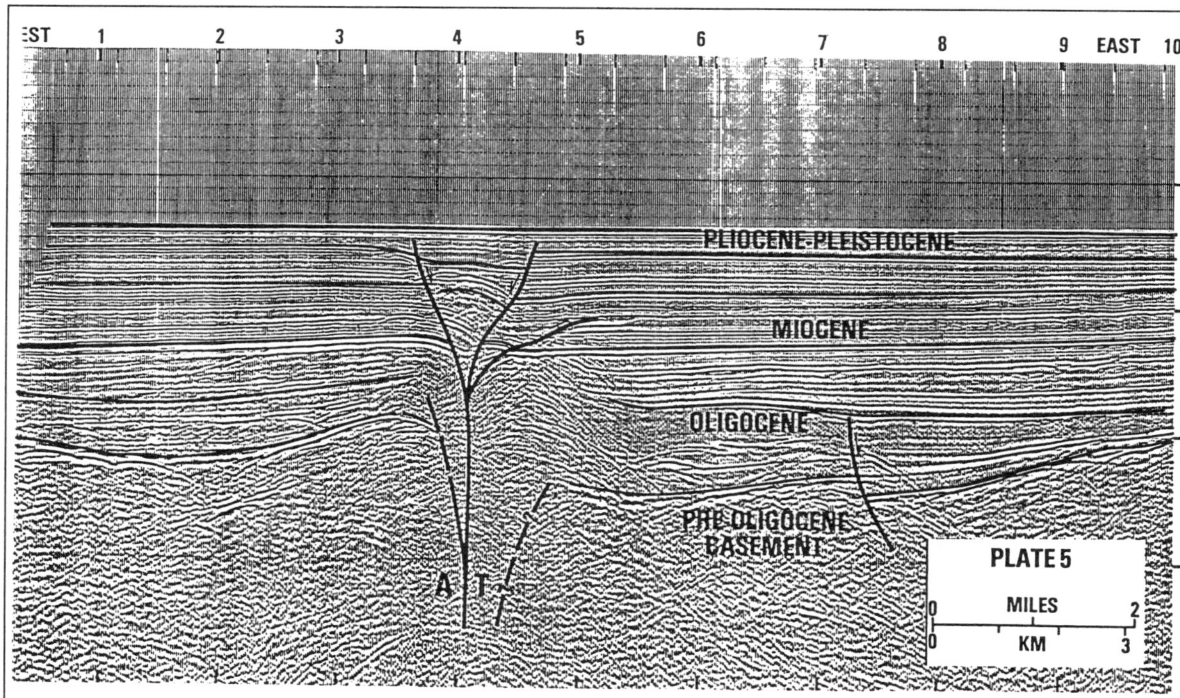

Wrench Fault Tectonics 4.2-7

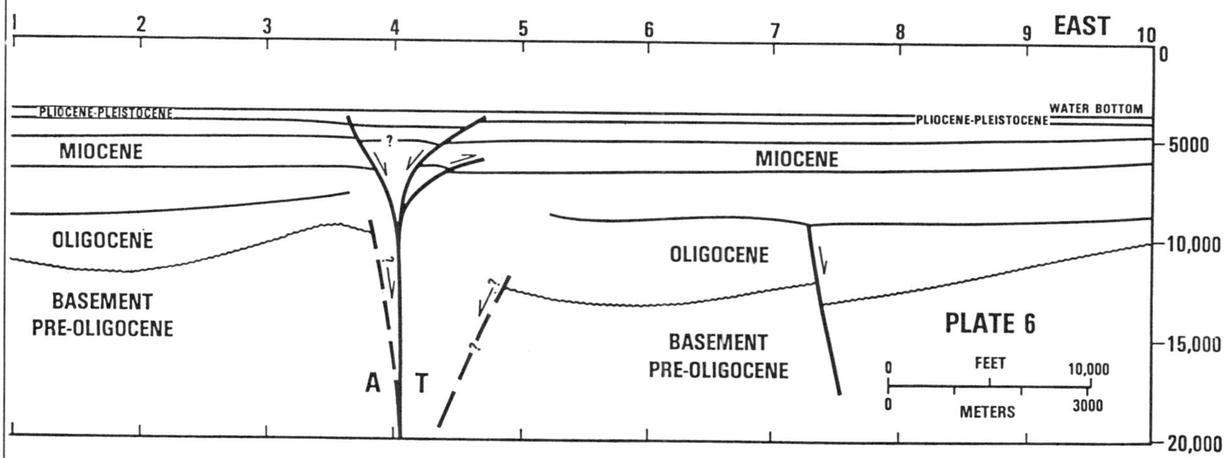

Wrench Fault Tectonics 4.2-8

Fig. I-12/C - Flower-structures - Negative - *Structures en fleurs - Négatives*
(HARDING, 1983) *(HARDING, 1983)*

"EN-ECHELON" EXTENSIONAL FRACTURES (MATTAUER)

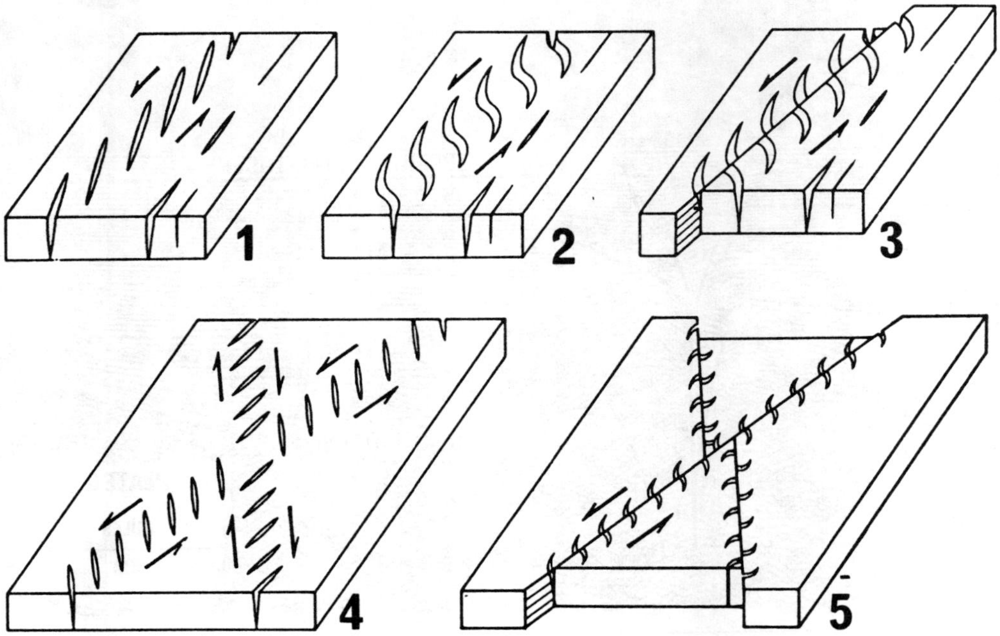

Generated by lateral shifting between neighbouring blocks.

1 - straight 2 - twisted 3 - faulted fractures
and 4 - straight and 5 - twisted and faulted fracture-networks.

fentes en échelons

Divers aspects de fentes droites (1), tordues (2), faillées (3), et réseaux de fentes en échelons conjugués, droites (4), tordues et faillées (5)

Fig. I-13 - Open fractures (MATTAUER)

Fig. I-13 - Fractures ouvertes (MATTAUER)

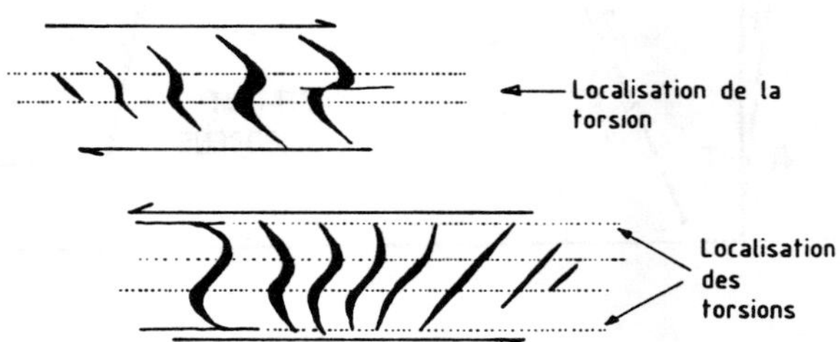

Evolution of extensional (open) fractures (J. HENRY)

Evolution des fentes de tension sigmoïdes (J. HENRY)

Fig. I-13/B

Fig. I-13/B

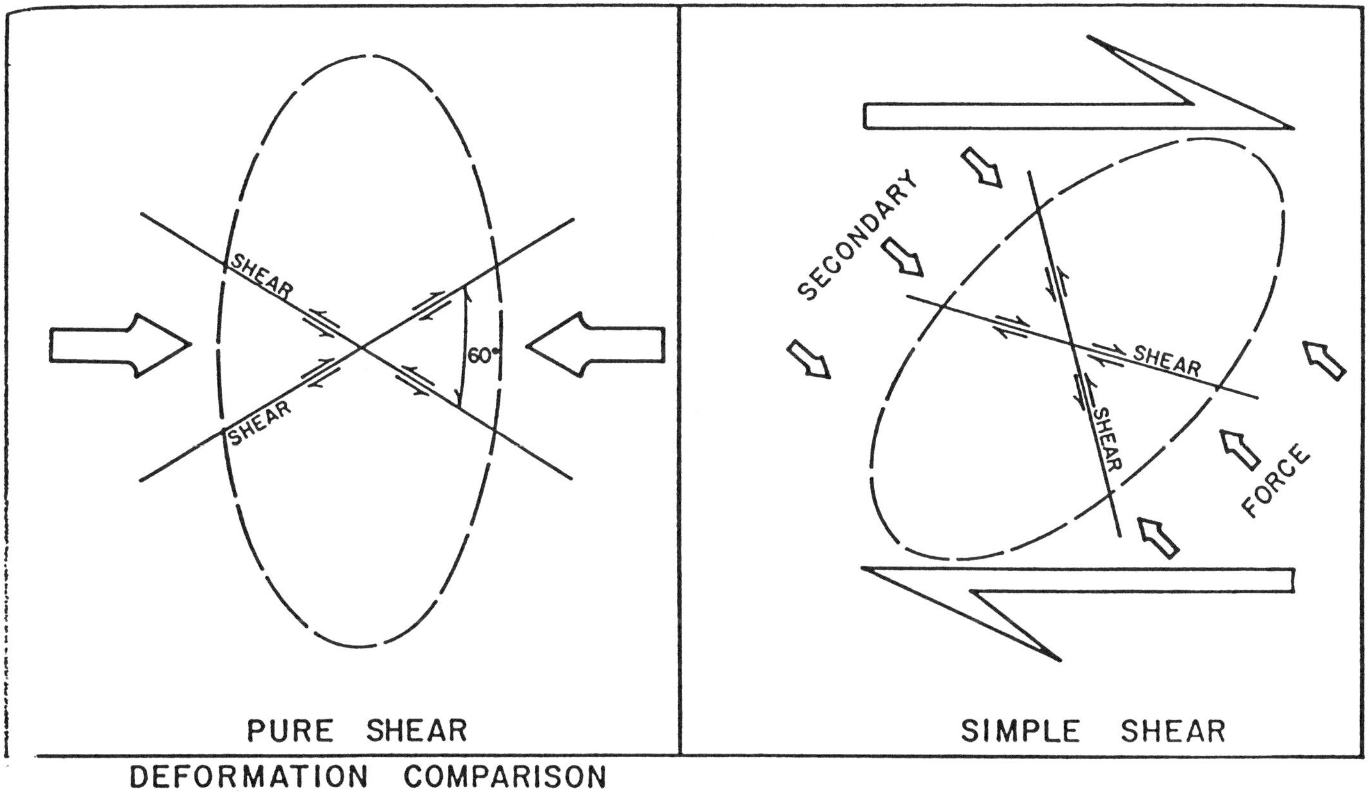

Fig. I-14 - Pure shear vs. simple shear
(THOMAS, 1971)

- *Cisaillement pur et cisaillement simple*
(THOMAS, 1971)

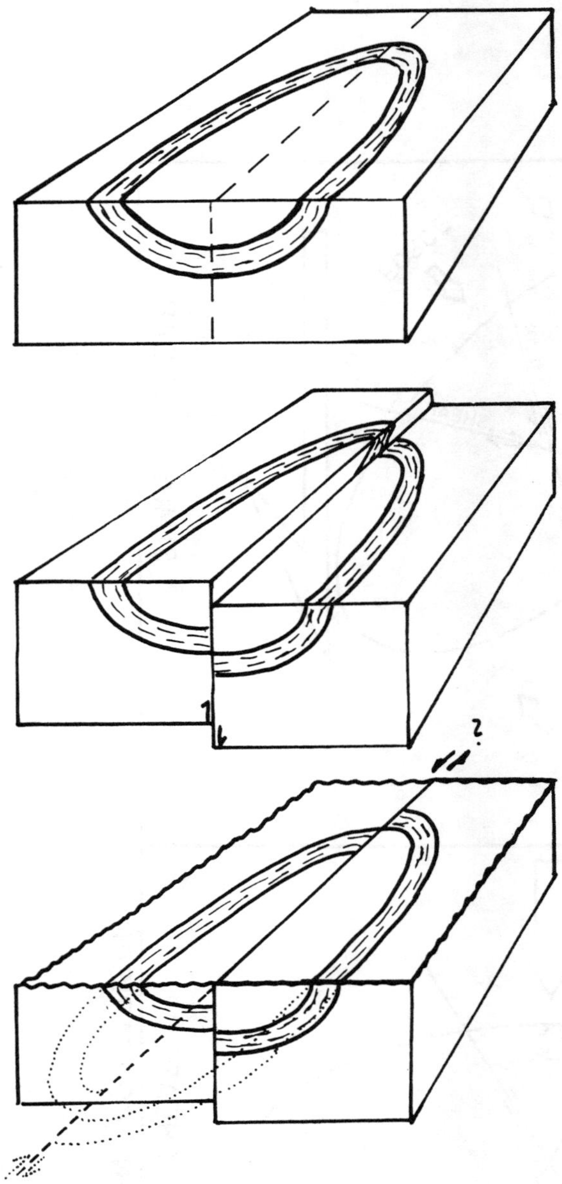

DECROCHEMENT APPARENT
(observer l'autre extrémité de la structure !...)

"FALSE" WRENCH-MOVEMENT

(observe the other end of the structure!...)

Fig. I-15 - What is <u>not</u> a wrench-fault ? (G.Z.)

- *"Ce qui n'est pas un décrochement"* (G.Z.)

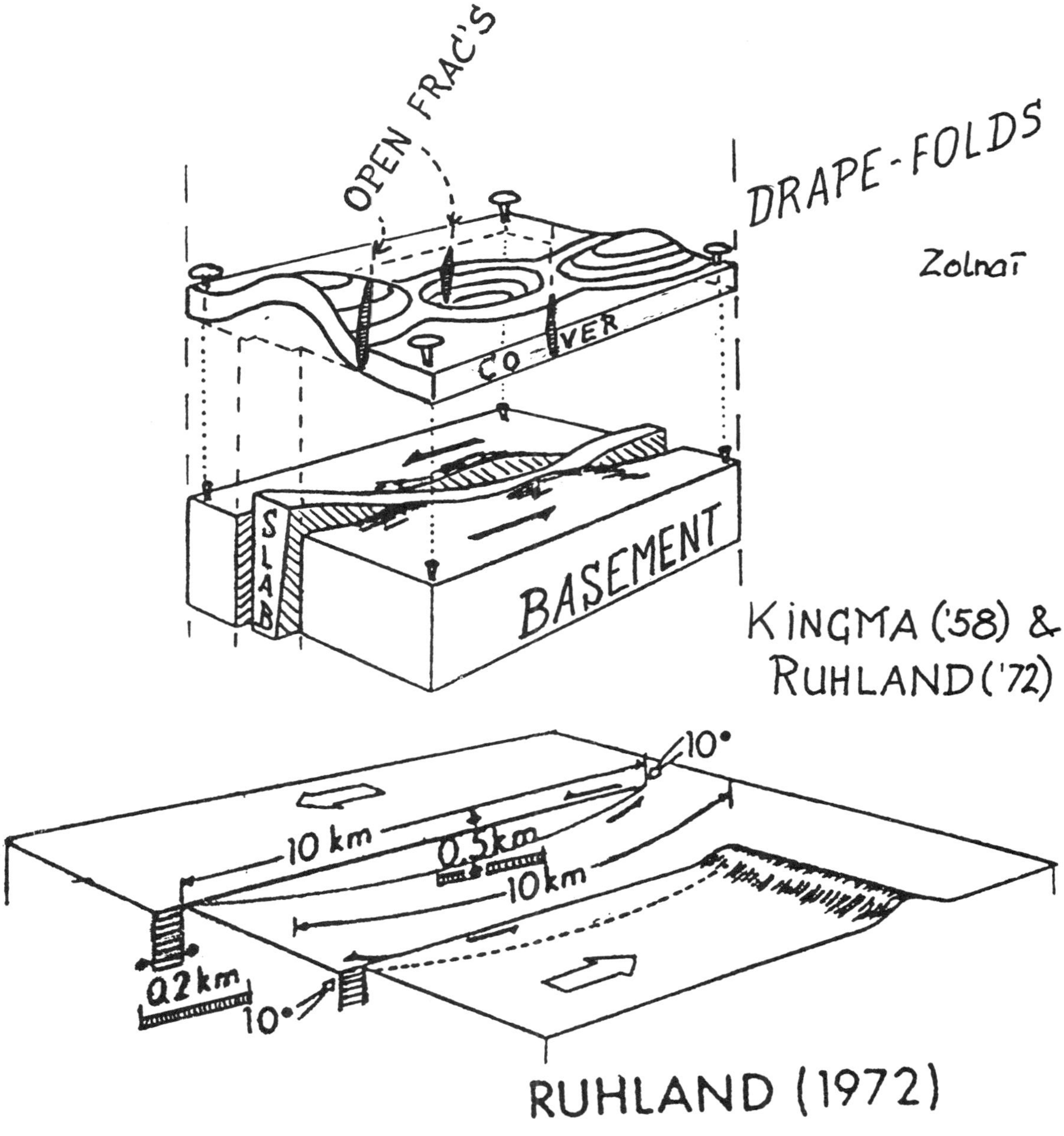

Fig. I-16 - Drape folds (G.Z.)

Fig. I-16 - Plis de drapage (G.Z.)

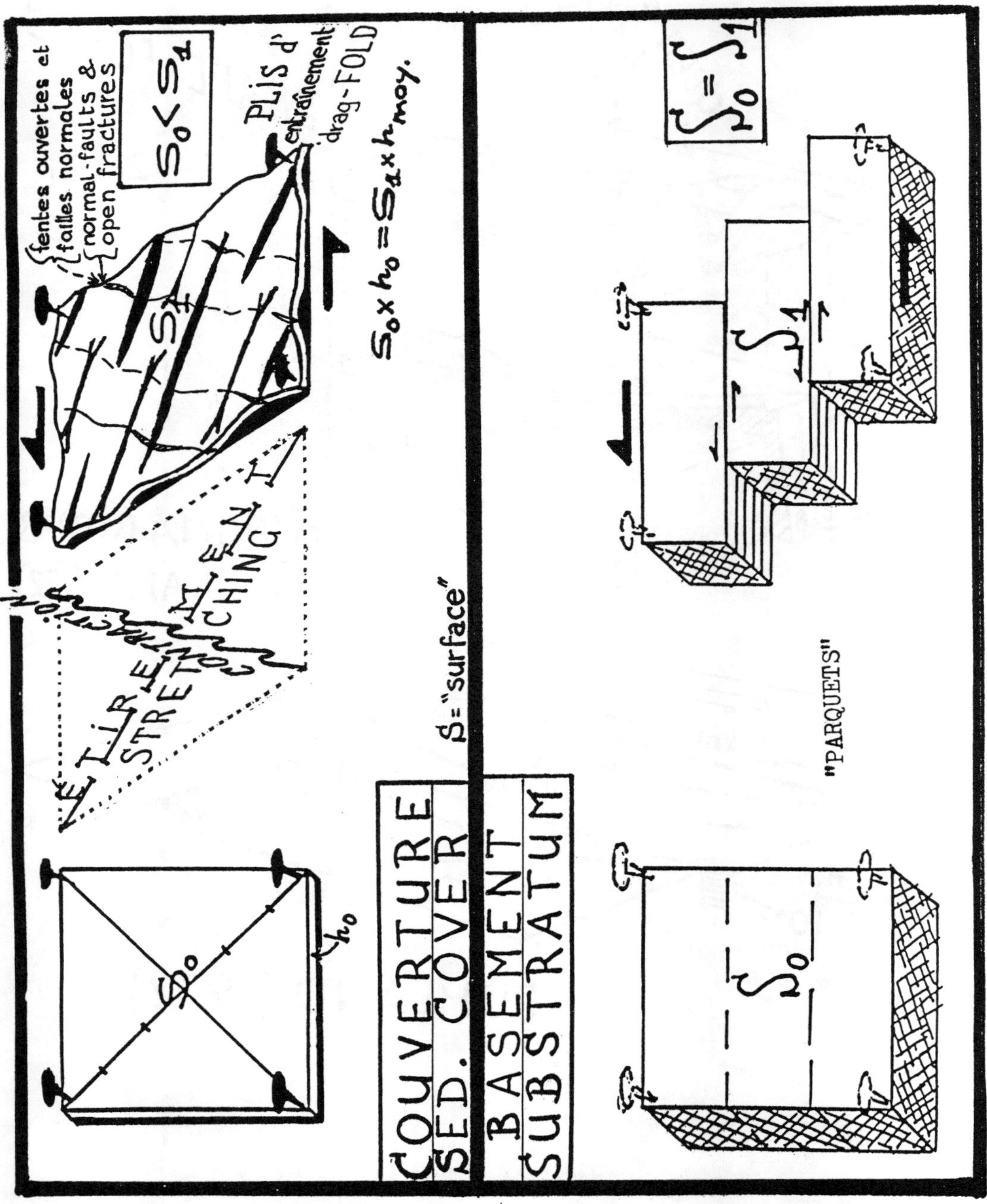

Fig. I-17 - Drag-folds
(G.Z.)

Fig. I-17 - Plis d'entraînement
(G.Z.)

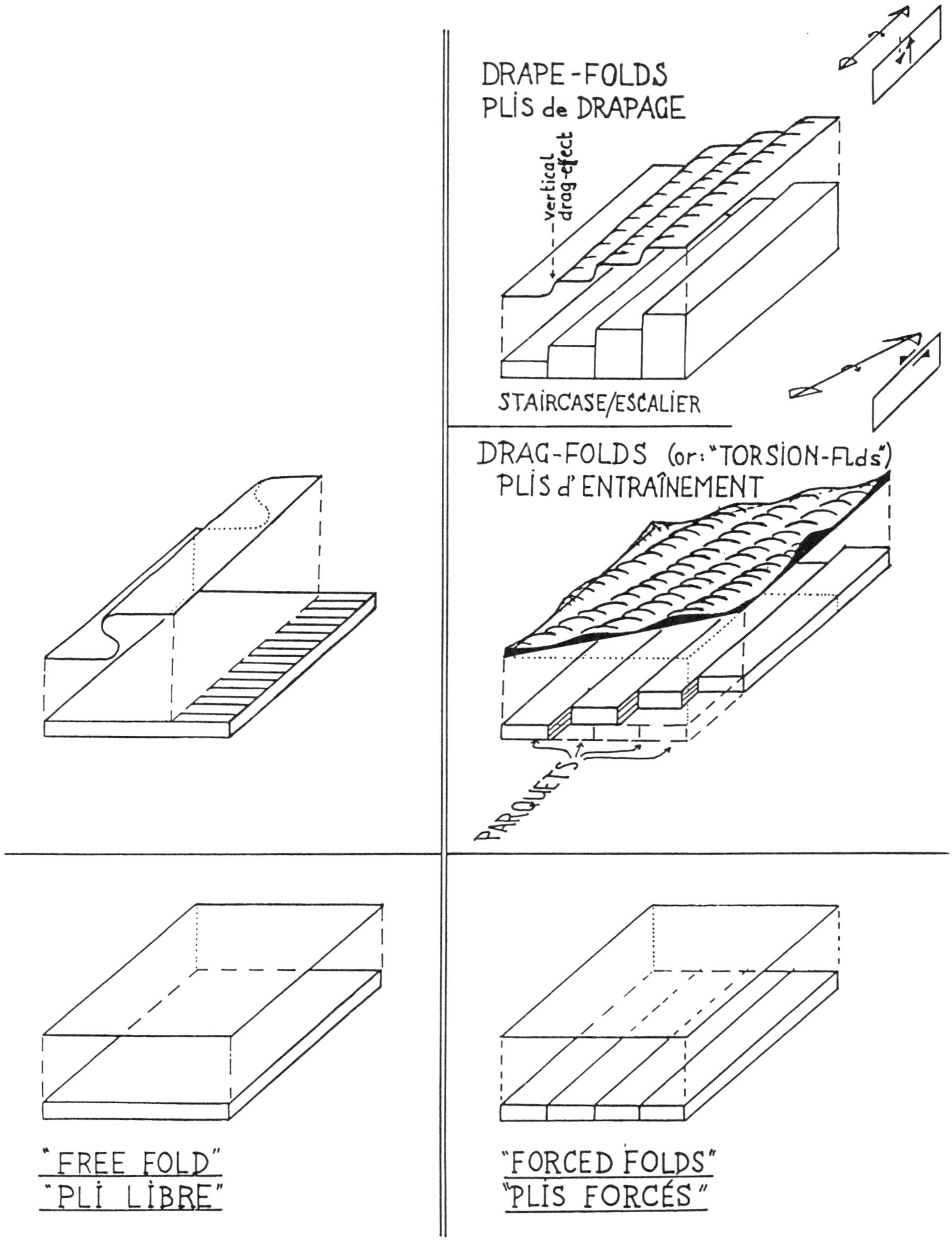

Fig. I- 18. FREE FOLDS - FORCED FOLDS
PLIS LIBRES - PLIS FORCES

FIGURES

II. WRENCHING VS. BLOCK-FAULTING

DECROCHEMENTS ET SYSTEMES DE FAILLES

II-B) FIGURES

II-B) FIGURES

		Fig. n°
. Couloirs (bandes) de cisaillement et systèmes de fracture Montana central (USA)	/JOHNSON/	II-1
Bourgogne (France)	/BERGERAT/	II-2
. Zones de cisaillements majeures, linéaments	/BAARS/	II-3
. Bassins losangiques	/DE VOTO, PATTERSON/	II-4
	/HAM, WALPER/	II-5
. Blocs hauts et effondrements associés	/CROWELL/	II-6
. Touches de piano	/DE WIT/	II-7
	/G.Z./	II-8
. Systèmes de fracture du Montana Central	/U.S.G.S./	II-9

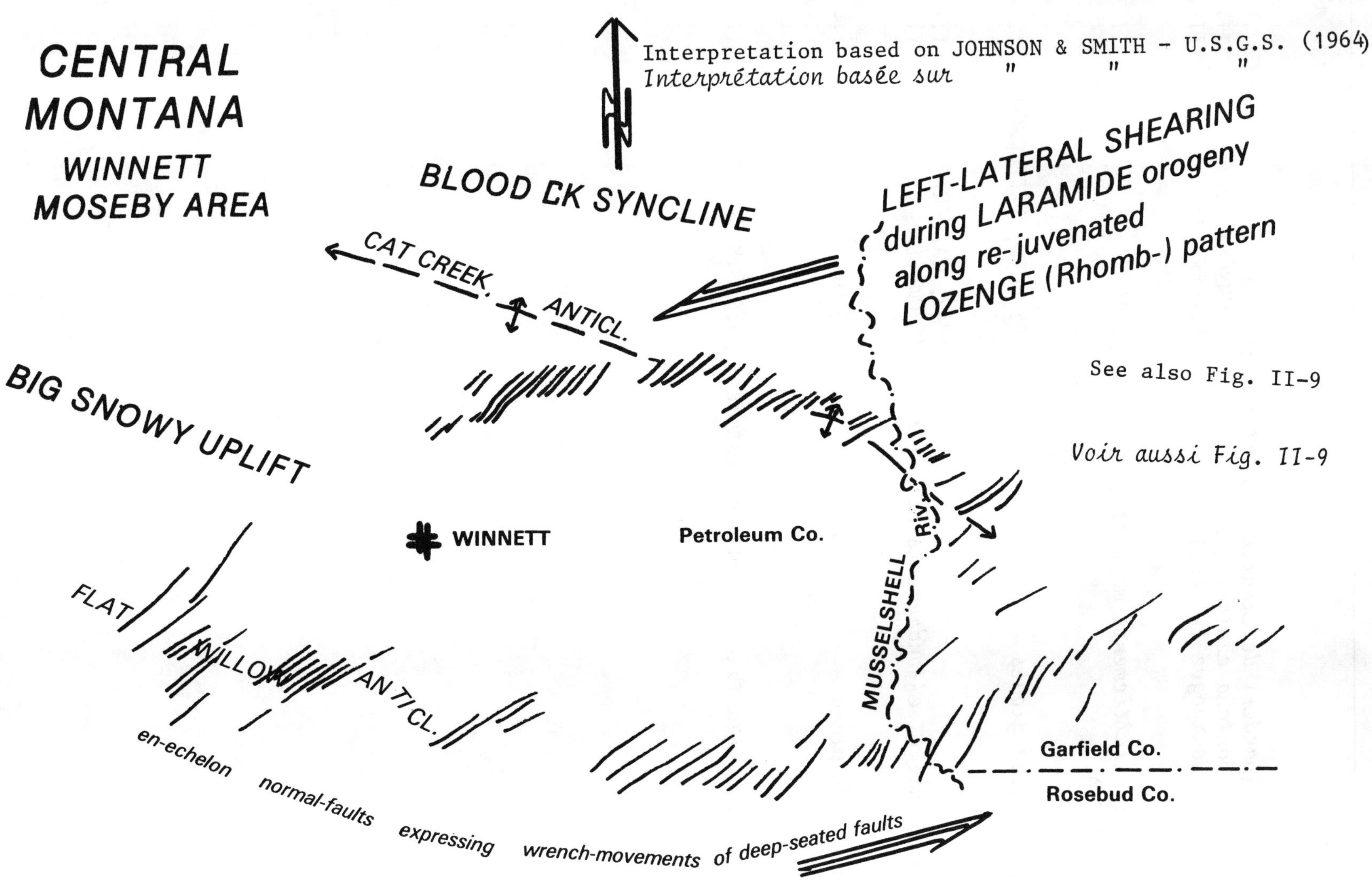

6 mi = 10 km

Fig. II-1 - Wrench-corridors and Gash-faults : Central Montana (USA)

Couloirs (bandes) de cisaillement et systèmes de fracture : Montana central (USA)

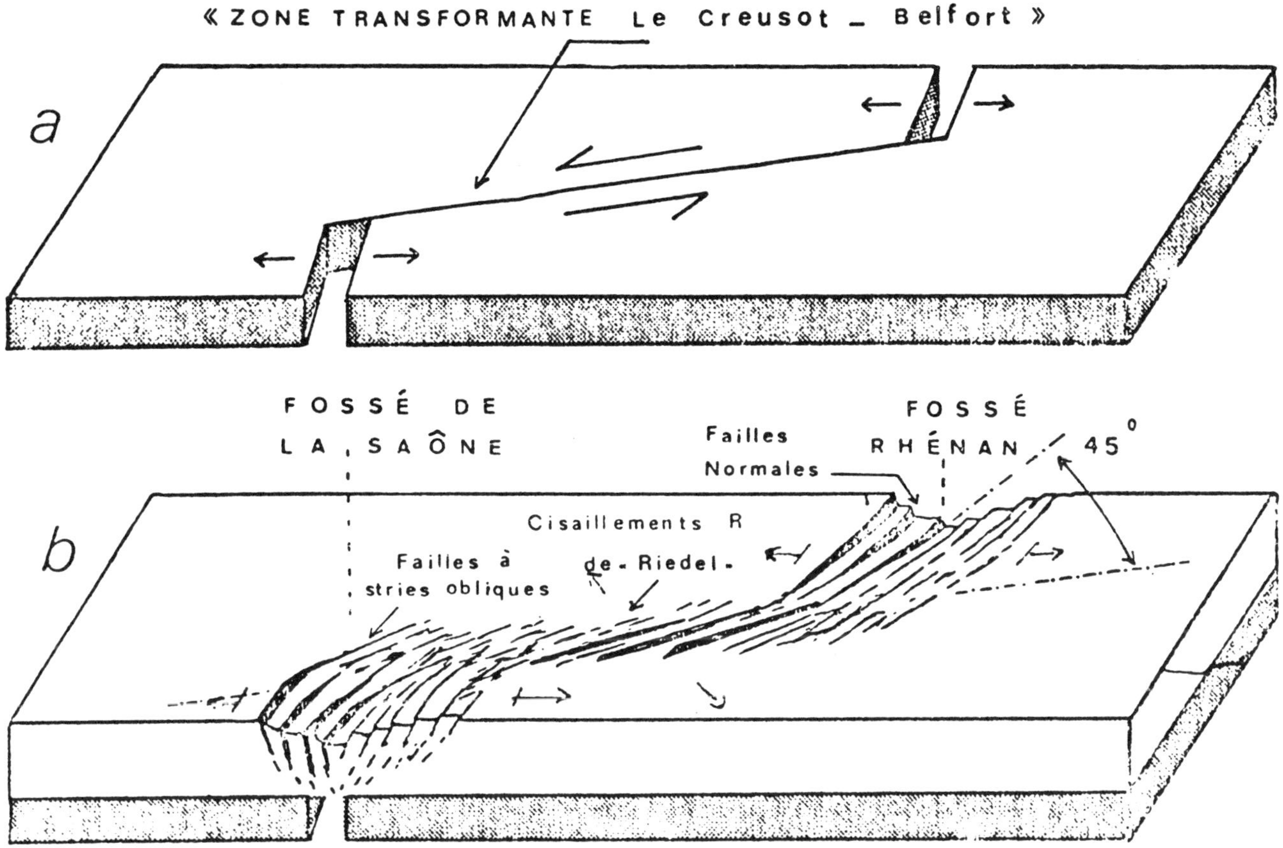

FIGURE 6

Liaison de deux zones d'extension — le Fossé de la Saône et le Fossé Rhénan — par une « zone transformante Le Creusot-Belfort ».

Connection of two extension zones — saonish and rhenish troughs — by a "transform zone Le Creusot-Belfort". Application du dispositif expérimental de P. Tapponnier et J. Varet (1974). Schématisation des structures observées.

Fig. II-2 - Wrench-corridors and Gash-faults : Bourgogne (France) (BERGERAT)

Fig. II-2 - Couloirs (bandes) de cisaillement et systèmes de fracture : Bourgogne (France) (BERGERAT)

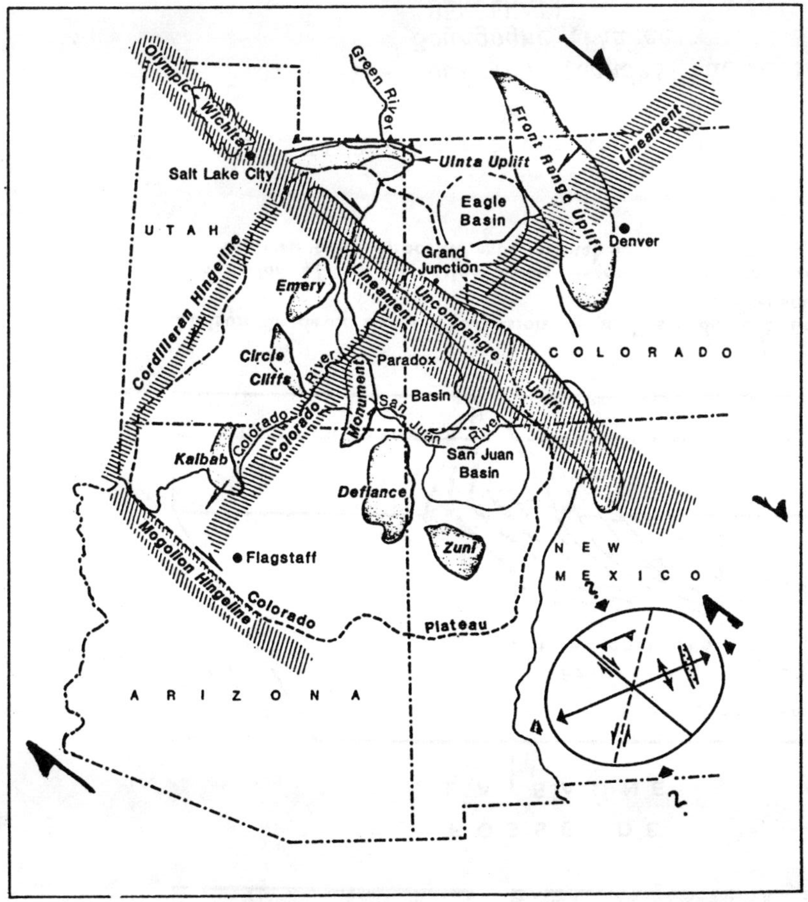

—Map showing location of Colorado Plateau and relationship to major orthogonal set of basement lineaments. Northwesterly lineaments are right lateral, northeasterly lineaments are left lateral. Stress-strain ellipsoid is so oriented that maximum compressive stress is directed from the north.

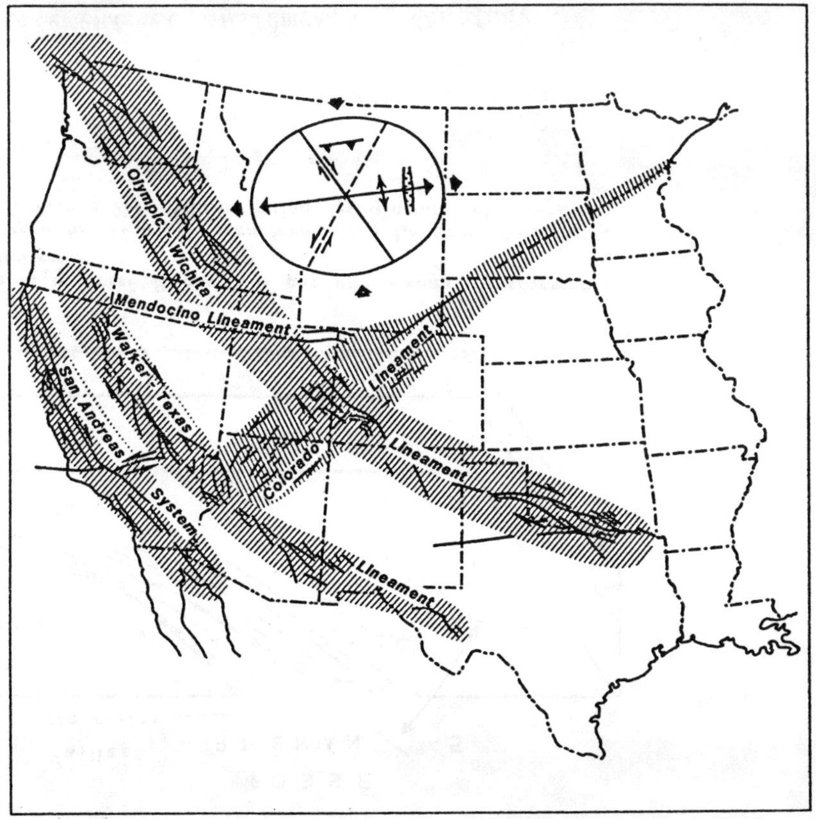

—Map showing locations of major basement lineaments of western United States. Sense of strike-slip offset is shown by arrows; northwesterly lineaments are right lateral, northeasterly lineaments are left lateral. Stress-strain ellipsoid is so oriented that maximum compressive stress is directed from the north.

Fig. II-3 - Mega-shears and lineaments (BAARS)

Fig. II-3 - Zones de cisaillements majeures, linéaments (BAARS)

Fig. II-4 - Lozange-basins
(DE VOTO, PATTERSON)

Fig. II-4 - Bassins losangiques
(DE VOTO, PATTERSON)

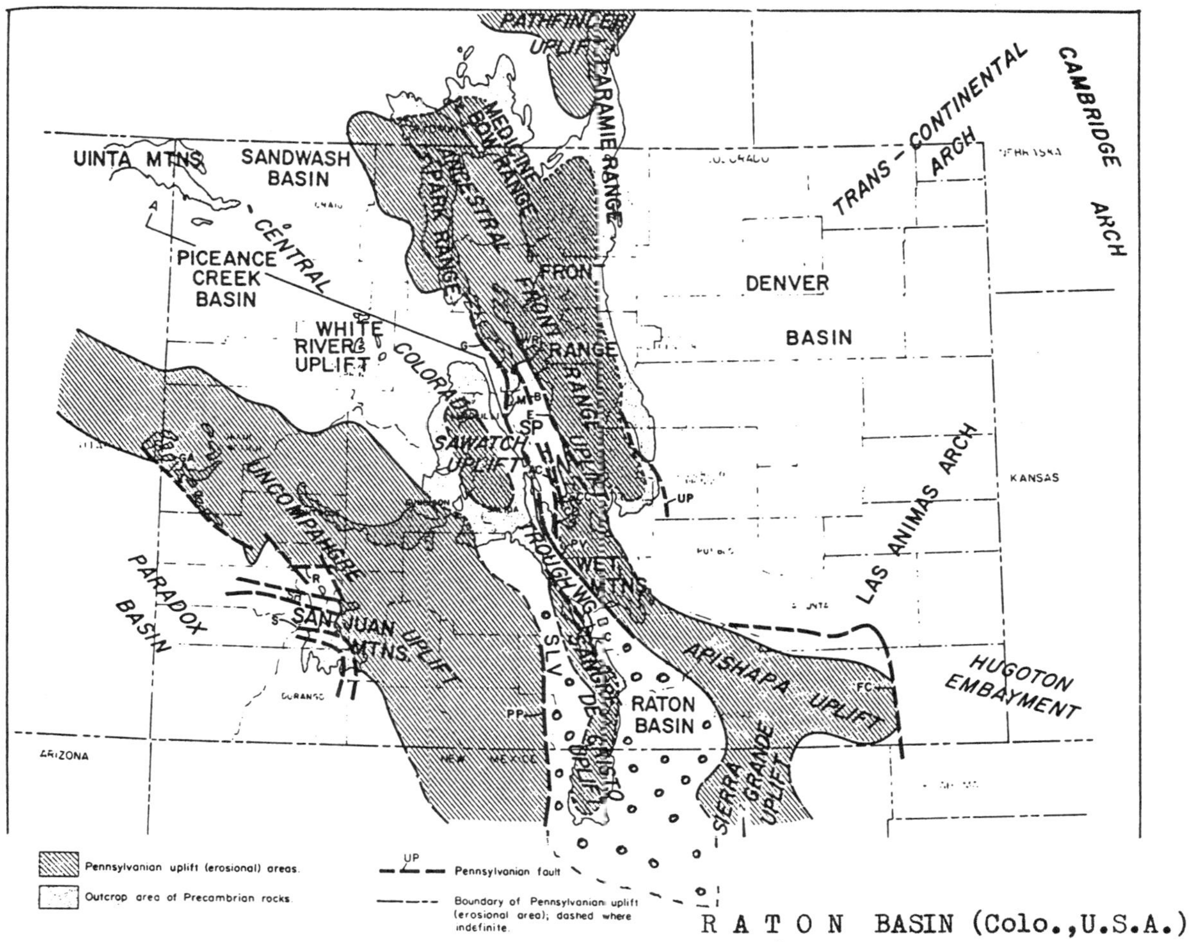

R A T O N BASIN (Colo.,U.S.A.)
From/de: R.de VOTO - 1980

Pennsylvanian and Cenozoic tectonic features. Pennsylvanian features shown with inclined lettering; Cenozoic features, with vertical lettering. Sawatch and Sangre de Cristo uplifts interpreted as Pennsylvanian tectonic features

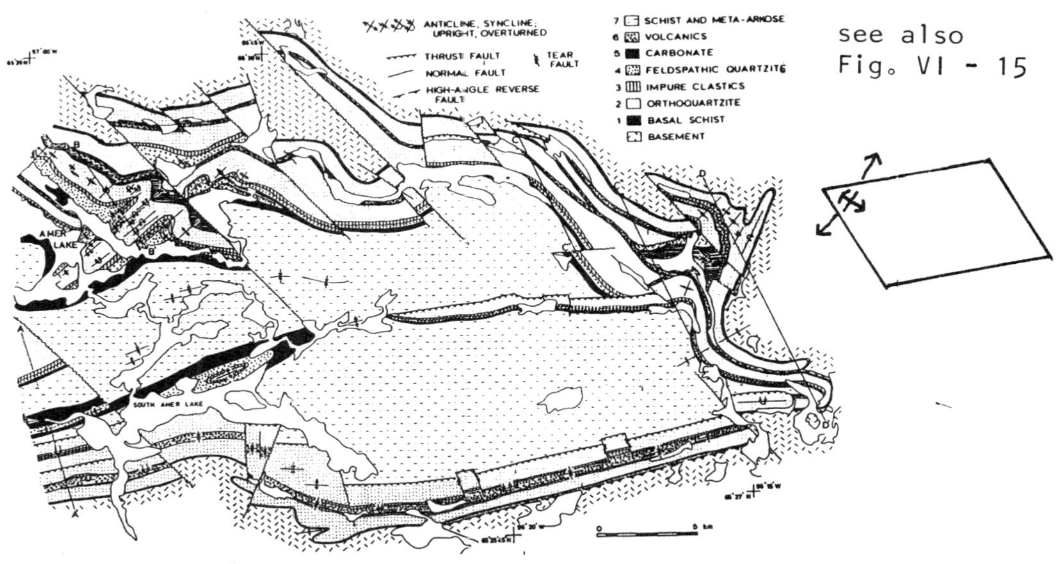

see also
Fig. VI - 15

Geological map of the eastern end of the Amer Belt.

A M E R Belt, N.W.Canada. From/de: PATTERSON - 1986

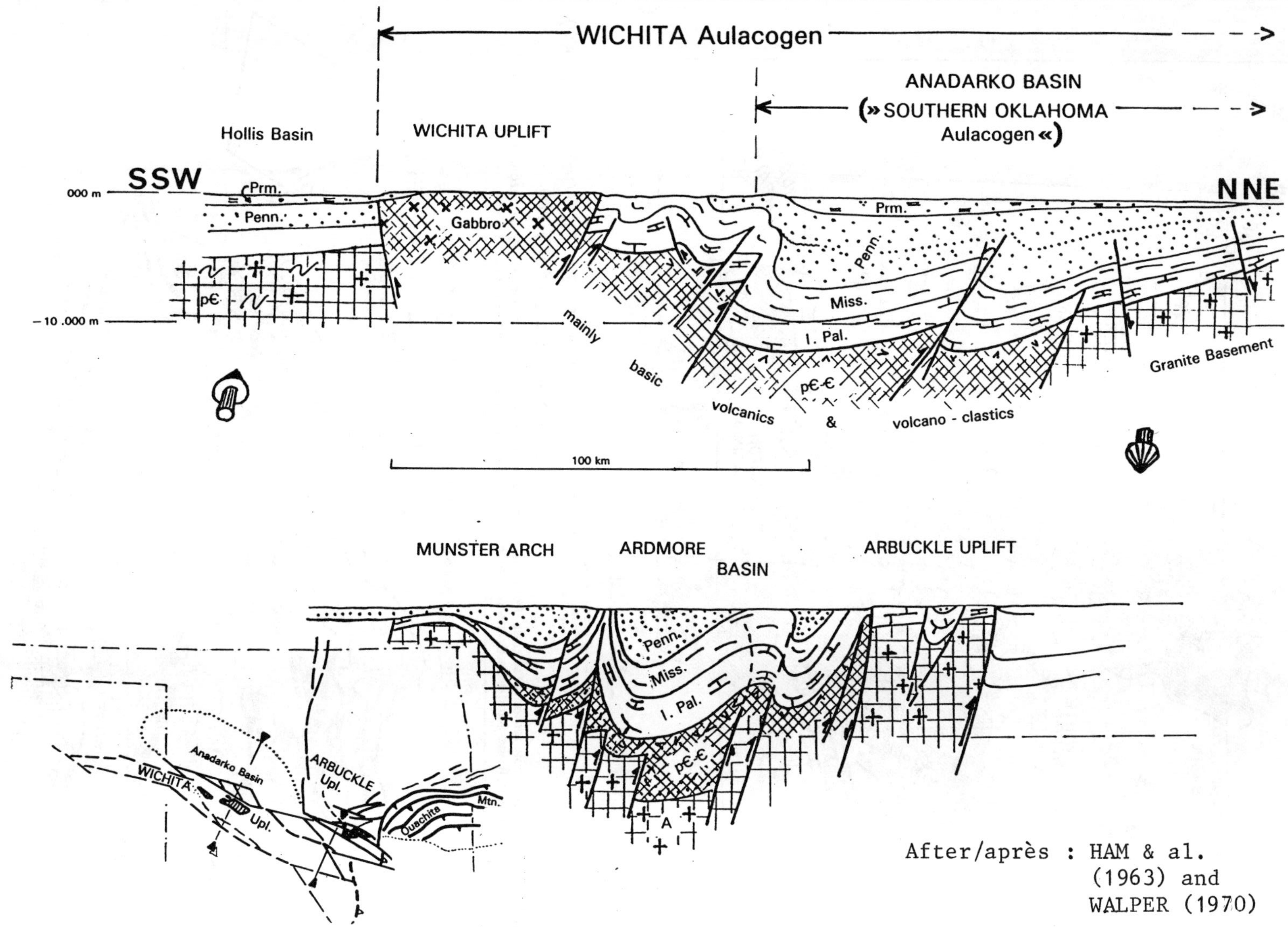

After/après : HAM & al. (1963) and WALPER (1970)

Fig. II-5/A - Paired uplifts and basins
Blocs hauts et effondrements associés
(Région : OKLAHOMA - TEXAS)

From HAM & al. (1963)

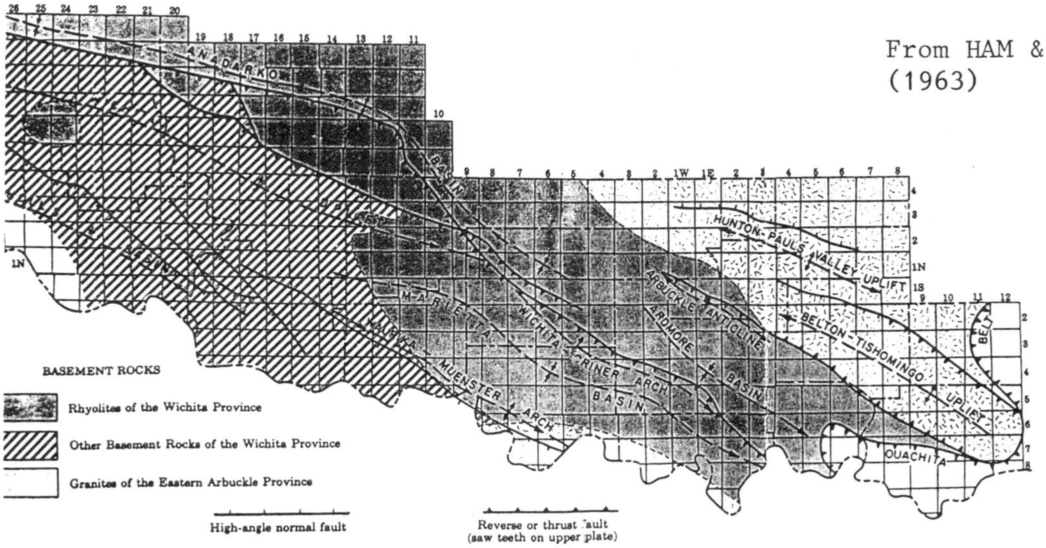

INDEX MAP SHOWING RELATION OF MAJOR STRUCTURAL FEATURES TO MAJOR BASEMENT-ROCK GROUPS IN SOUTHERN OKLAHOMA

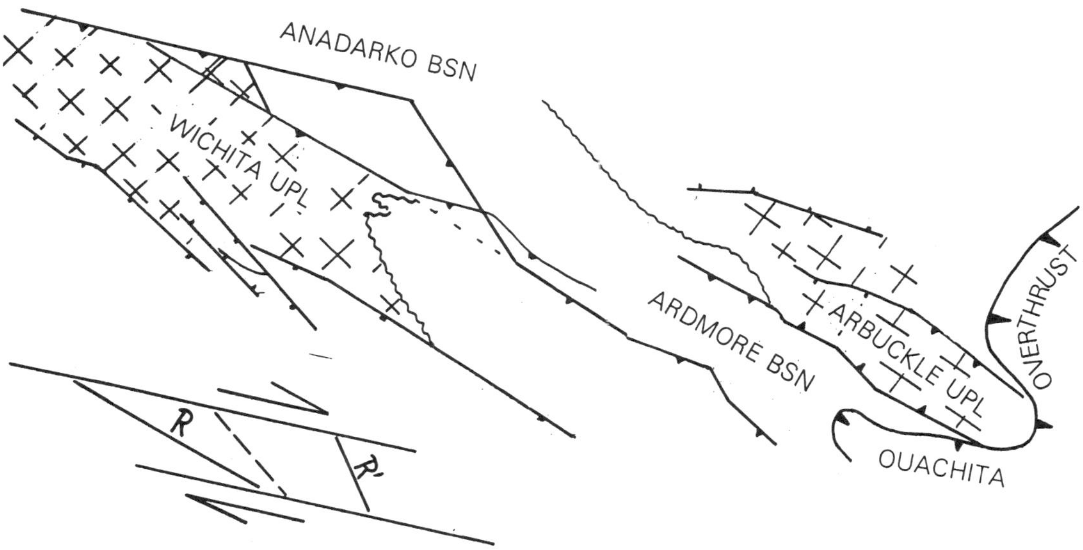

The WNW-ESE - oriented lozenge-pattern is here interpreted as being due to a right-lateral shear-couple, intervening during the late-Paleozoic ("Ancestral Rockies" - Ouachita - Appalachian) orogeny. See also Fig. VI-8 and VI-19.

Fig. II-5 B - Paired uplifts and basins (HAM, WALPER)

- Blocs hauts et effondrements associés (HAM, WALPER)

From CROWELL
(1974)

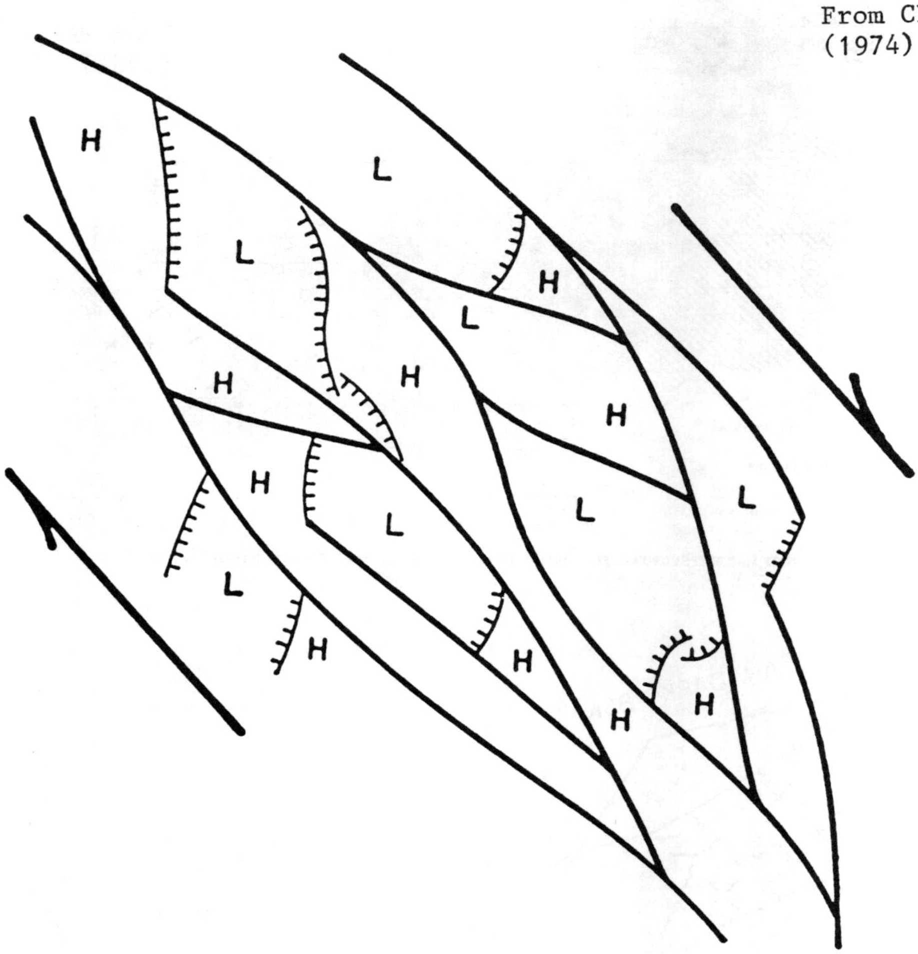

—Sketch map of region in a transform regime, showing pull-apart basins and tipped fault wedges where right-slip faults converge or diverge;

Fig. II-6 - Paired uplifts and basins
(CROWELL)

- *Blocs hauts et effondrements associés*
(CROWELL)

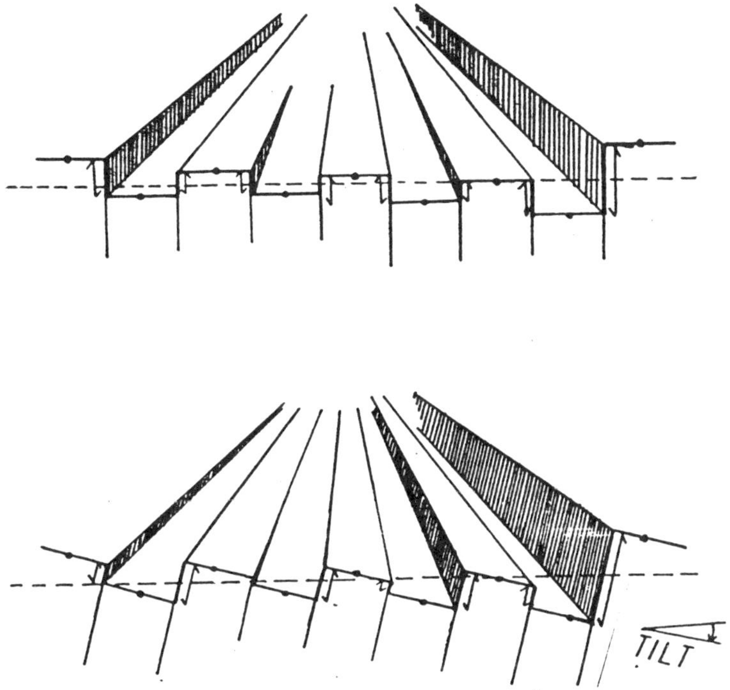

Fig. II-8 - Keyboards (touches de piano)
(G.Z.)
- *Touches de piano*

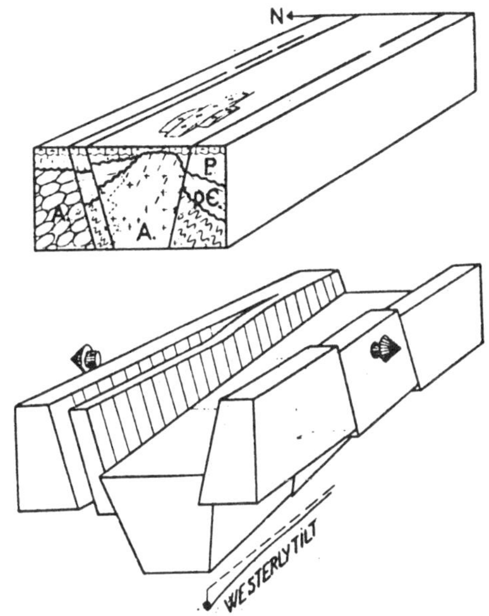

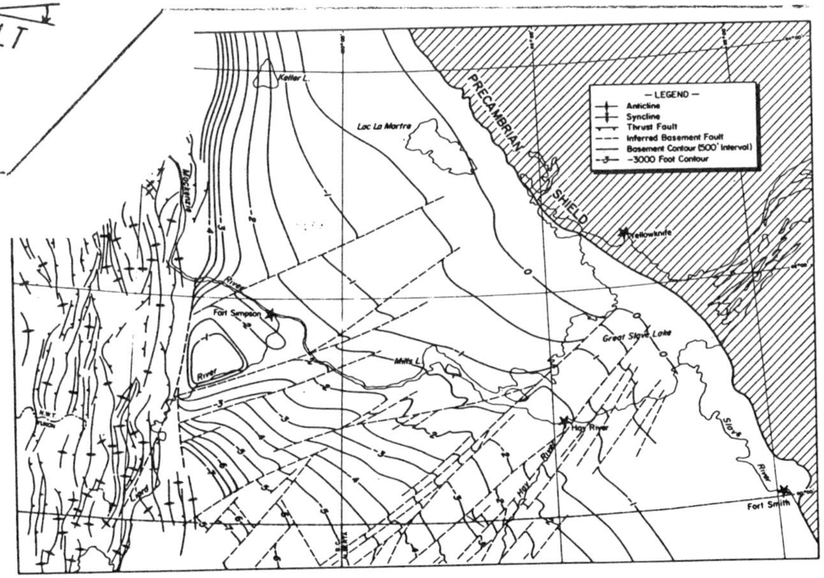

Fig. II-7 - Keyboards (touches de piano)
(DE WIT)

- *Touches de piano*

Area/Région :
Great slave lake - Liard plateau
(N.W. CANADA)

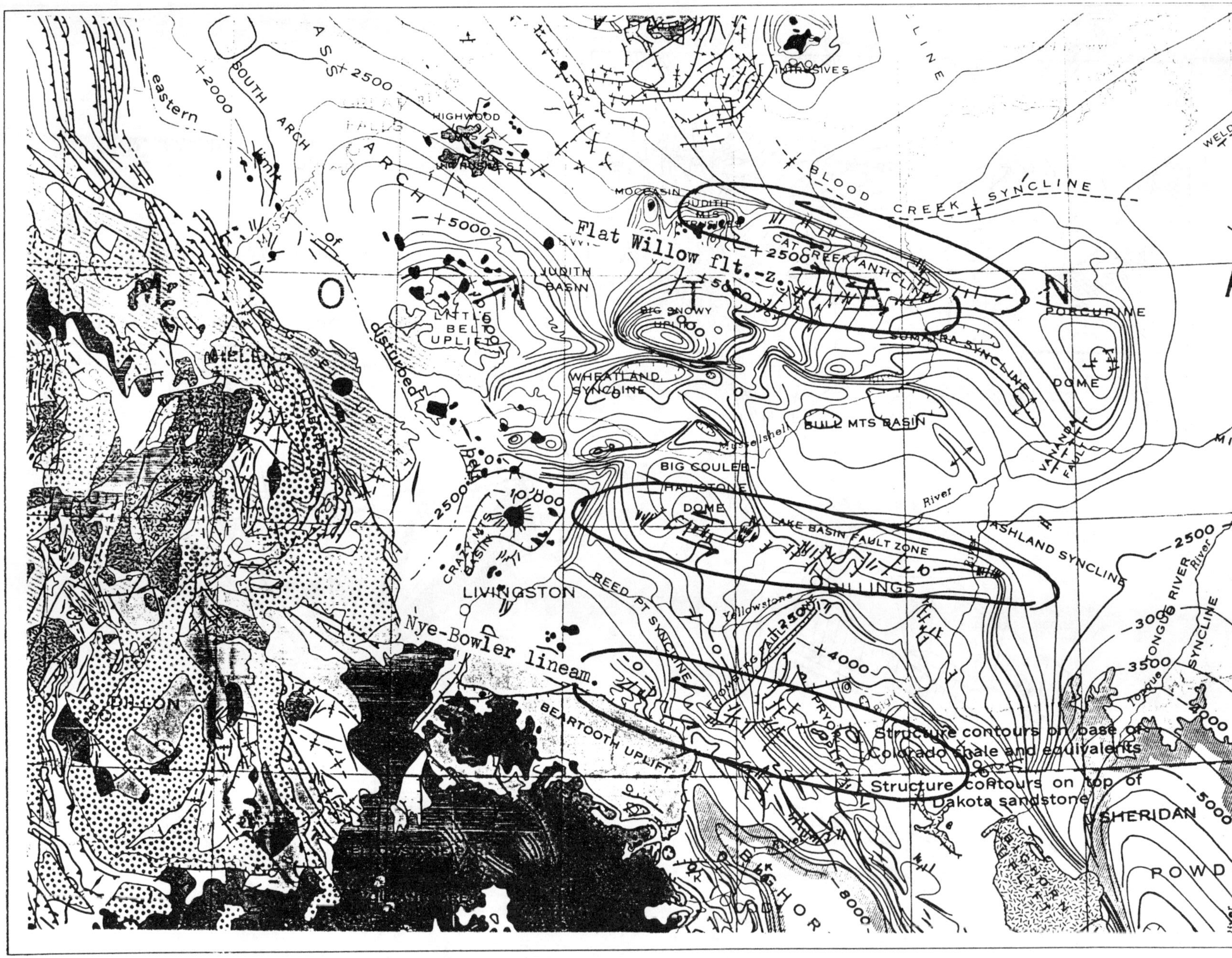

Fig. II-9 Gash-fault systems of Central Montana, U.S.A. (U.S.G.S.)

Système de fractures - Montana Central, U.S.A. (U.S.G.S.)

FIGURES

III. WRENCHING VS. FOLDING

DECROCHEMENTS ET PLIS

III-B) FIGURES

III-B) FIGURES

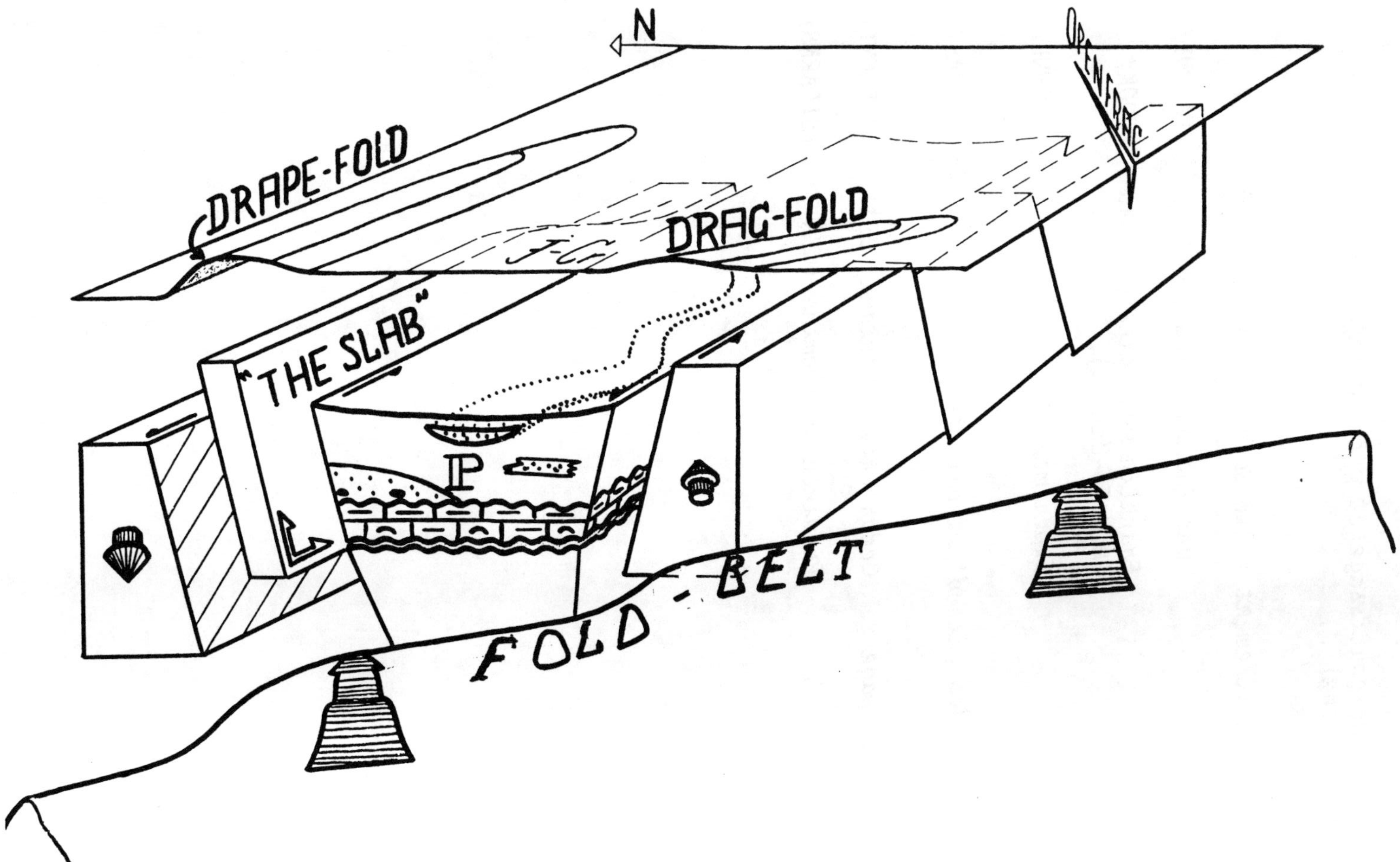

Fig. III-1 - Wrench-systems versus superficial folds (Ctr. Montana) (G.Z.)

- *Systèmes de coulissage et plis superficiels (Ctr. Montana) (G.Z.)*

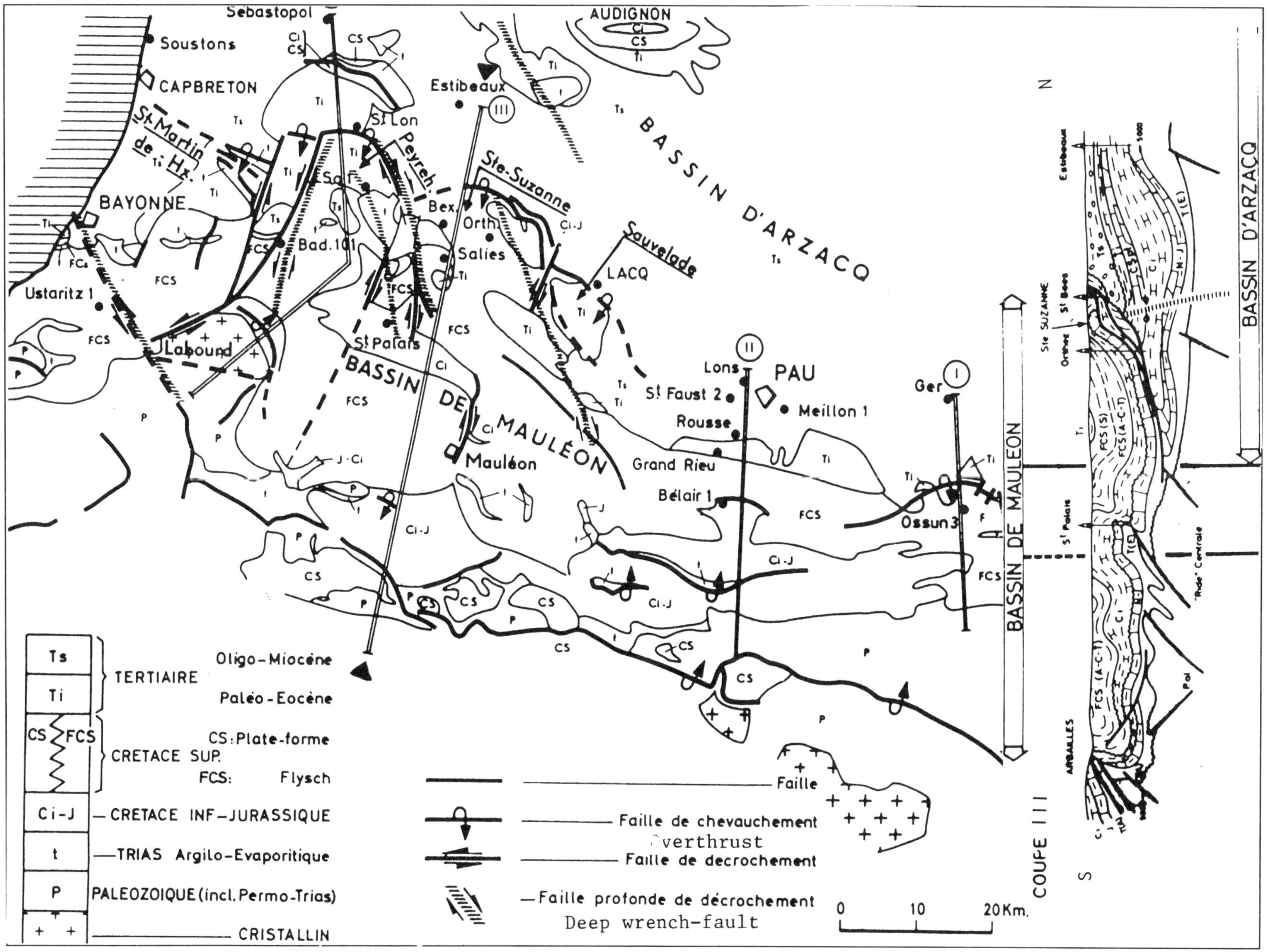

Fig. III-2 - Drag-folds : Ste Suzanne (s.w. France) (G.Z. 1971)

- *Plis d'entraînement : Ste Suzanne (France, S-O),* (G.Z 1971)

524 ASSEMBLÉE GÉNÉRALE DE LA S.G.S. À BIENNE

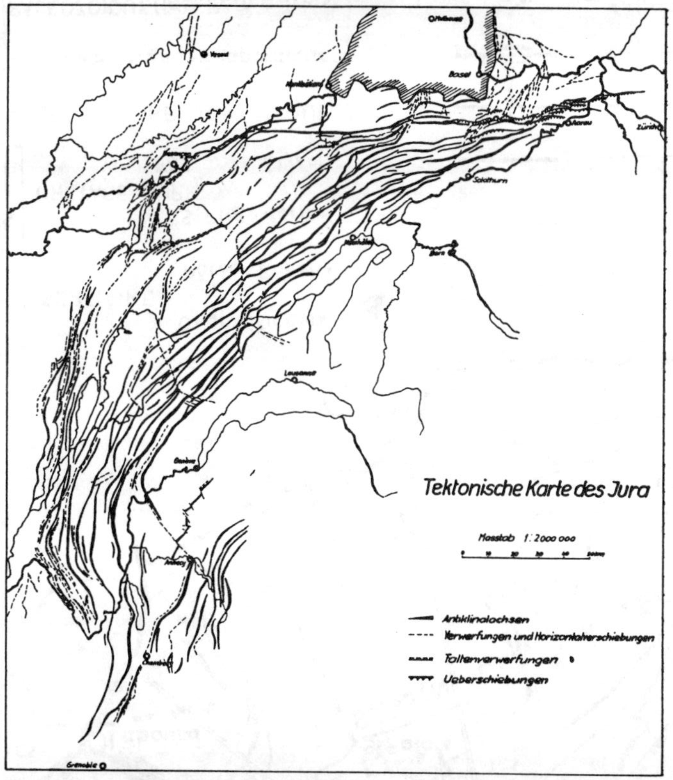

Fig. 4. Das Faltenbild des Faltenjura. Umgezeichnet und leicht ergänzt nach A. BERSIER (1934).

NAZARIO PAVONI: FALTUNG DURCH HORIZONTALVERSCHIEBUNG 525

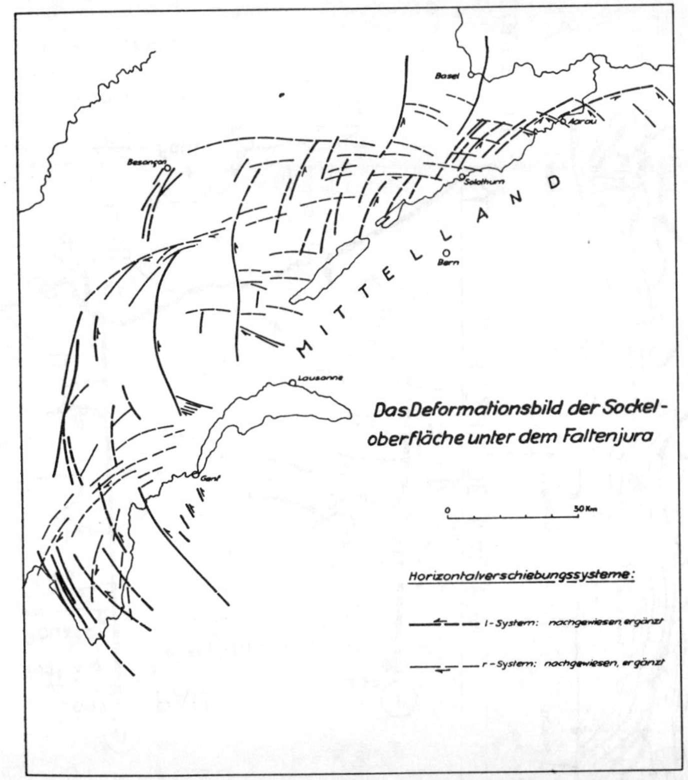

Fig. 5. Analyse des Faltenbildes des Faltenjura nach dem Prinzip der HV-Faltung, schematisierte und vereinfachte Darstellung. Ausgezogene Linien: Brüche und Horizontalverschiebungen an der Oberfläche sichtbar. Gestrichelte Linien: Horizontalverschiebungen abgeleitet aus dem Faltenbild.

TECTONIC MAP (surface-folds)

CARTE STRUCTURALE (plis de surface)

BASEMENT MAP (deep faults)

CARTE DU SOCLE (failles profondes)

Fig. III-3 - Drag-folds : Jura Mtns. (France and Switzerland)
(PAVONI)

- Plis d'entraînement : Jura franco-suisse,
(PAVONI)

PAVONI (1961)

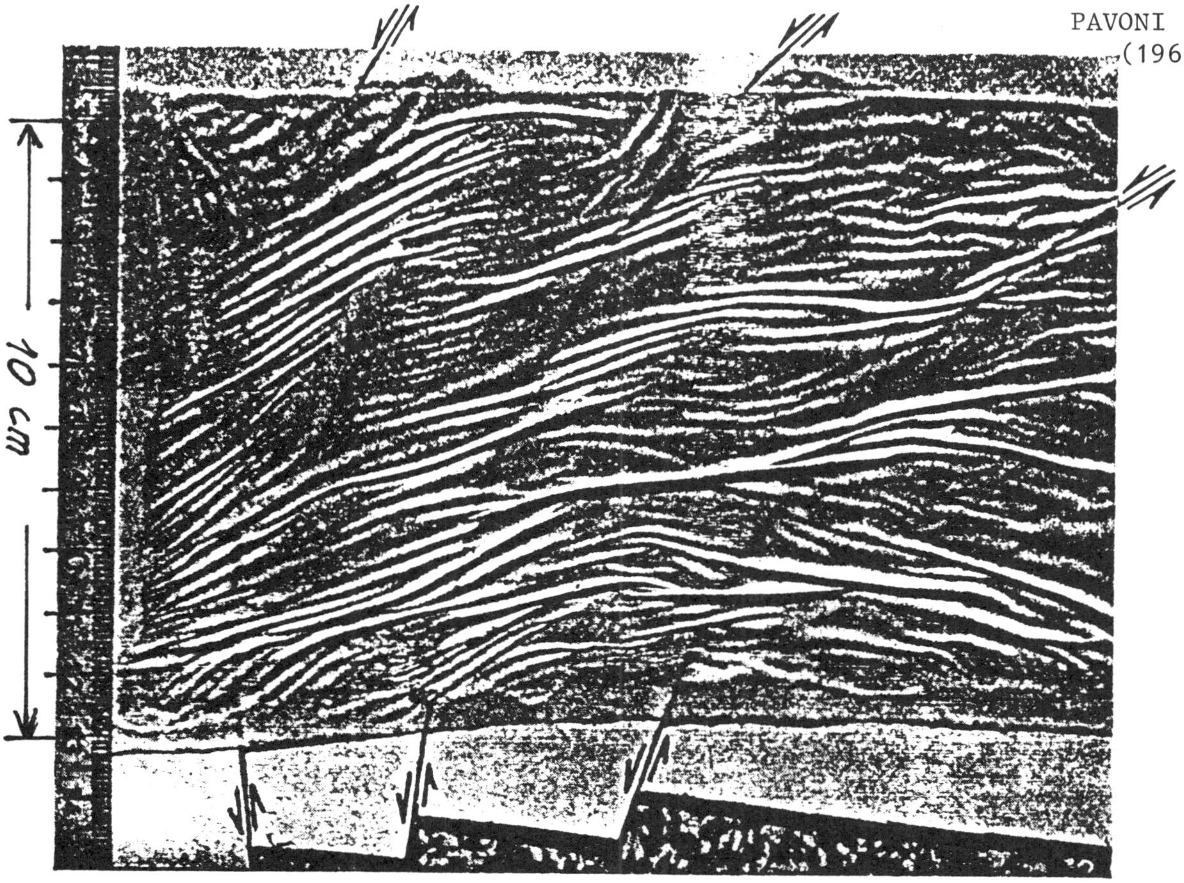

Faltung durch Horizontalverschiebung im Experiment. Hauchdünnes, nasses Papier über 0,5 mm mächtiger Schicht von Mehlpappe. Masstab in Zentimetern.

"FOLDING TROUGH WRENCHING" in experiment. Very thin, wet paper on 0.5 mm thick layer of papier mâché. - Scale in cm (to the left).

"PLISSEMENT PAR COULISSAGE HORIZONTAL" en expérience. Papier-pelure mouillé sur une couche de 0,5 mm de carton-pâte. - Echelle en cm (à gauche).

Fig. III-4 - Drag-folds : Jura Mtns. (France and Switzerland) (PAVONI)

- Plis d'entraînement : Jura franco-suisse, (PAVONI)

NORTHERN FRANKLIN MOUNTAINS — N.W.T. - N.W. CANADA

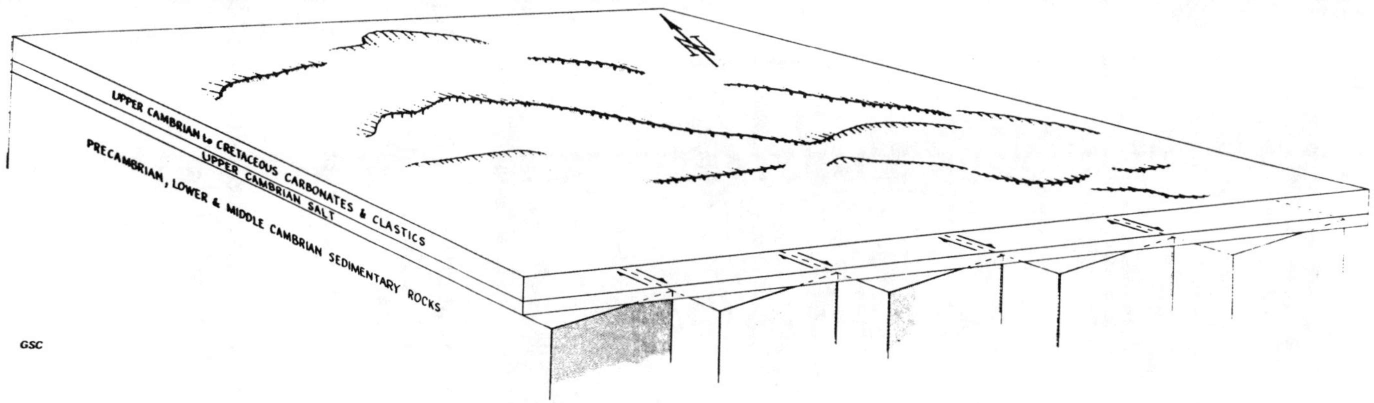

Schematic block diagram of northern Franklin Mountains showing Upper Cambrian salt as a ductile layer separating wrench-fault style below, from complex distributed-shear style above.

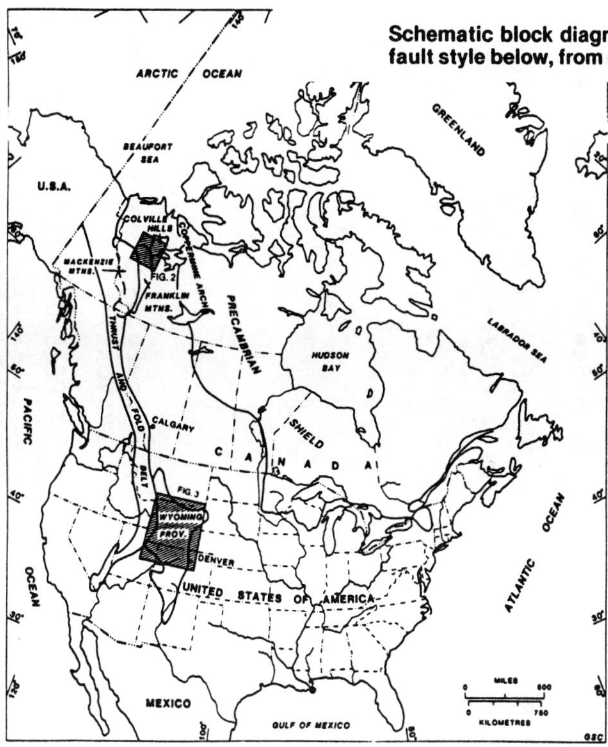

Figure 1. Location map showing Franklin Mountains relative to the Wyoming province.

Fig. III-5 - Drag-folds : Northern Franklin Mtns (N.W.T. Canada) (COOK)

- *Plis d'entraînement : Franklin Mountains du Nord (N.W.T.-Canada)* (COOK)

O D O N N E & V I A L O N (1983.)

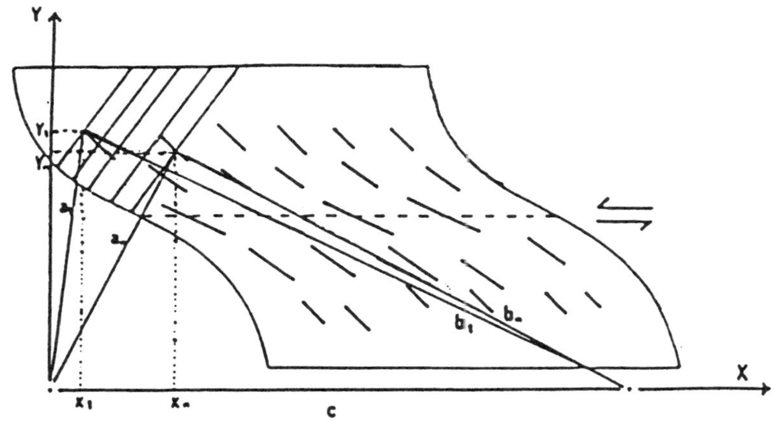

Triangulation of the crests of folds of a model. The $a_1 \ldots a_n$ and $b_1 \ldots b_n$ distances as well as the C base determine the (x_n, y_n) coordinates of the ends of each fold axis segment (between two cross-sections). Each segment is at an angle θ to the x-axis (parallel to the basement wrench fault).

F. O D O N N E (1981.)

Schéma synthétique par coupes séries dans un pli, montrant le changement de pendage du plan axial lorsque l'on franchit, en suivant le tracé de l'axe, la verticale du décrochement [d'après Odonne, 1981].

Synthetical sketch operating with consecutive cross-sections over a fold; it shows the change in dip of the axial plane when crossing over the wrench plane, following the axis of the fold.

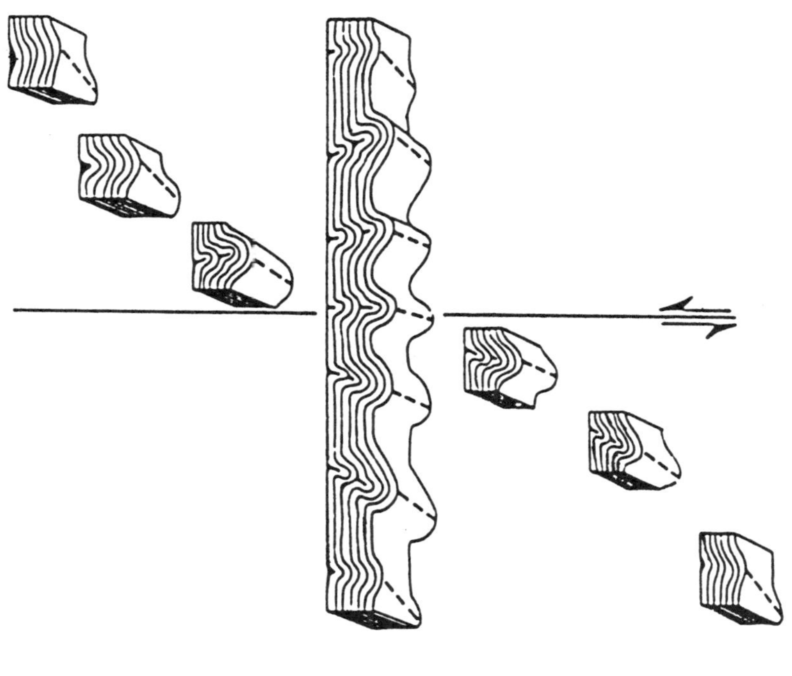

Fig. III-6 - Drag-folds : experimental (ODONNE)

- Plis d'entraînement : Experimental (ODONNE)

From : THOMAS (1971)

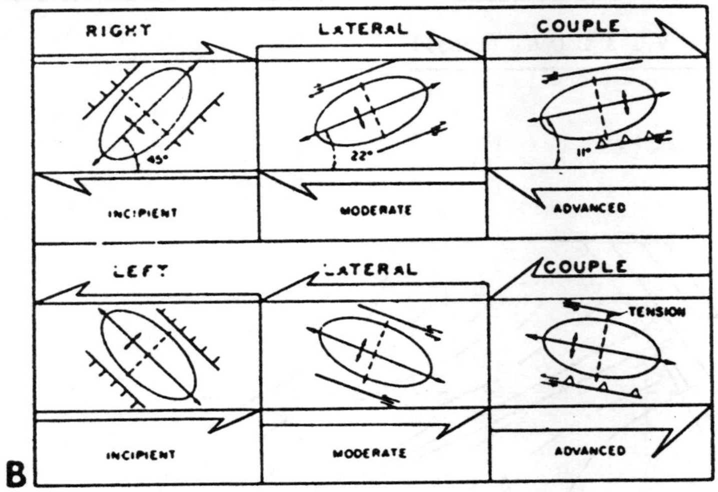

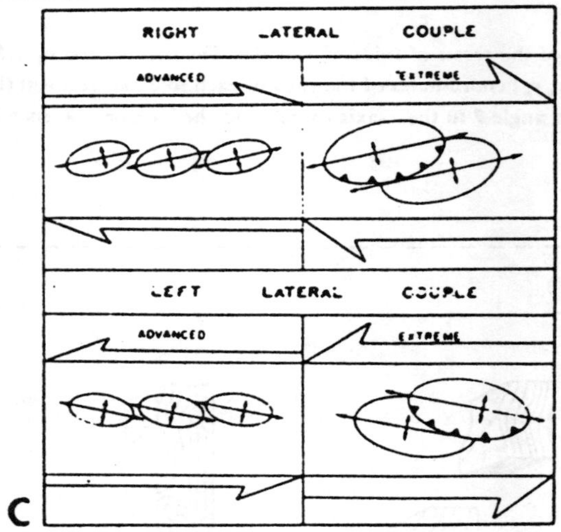

Various phases of simple shear mechanics (**B, C**).

B. Lineament simple shear produces block coupling which results in drag-fold uplifts with associated flank faults and crossfold tension. Phases of coupling classified as to degree of drag-fold rotation.

C. En echelon drag folds, characteristic of advanced phase, overthrust one another in extreme phase.

Fig. III-7 - Evolution of drag folds (THOMAS)

- *Evolution des plis d'entraînement (THOMAS)*

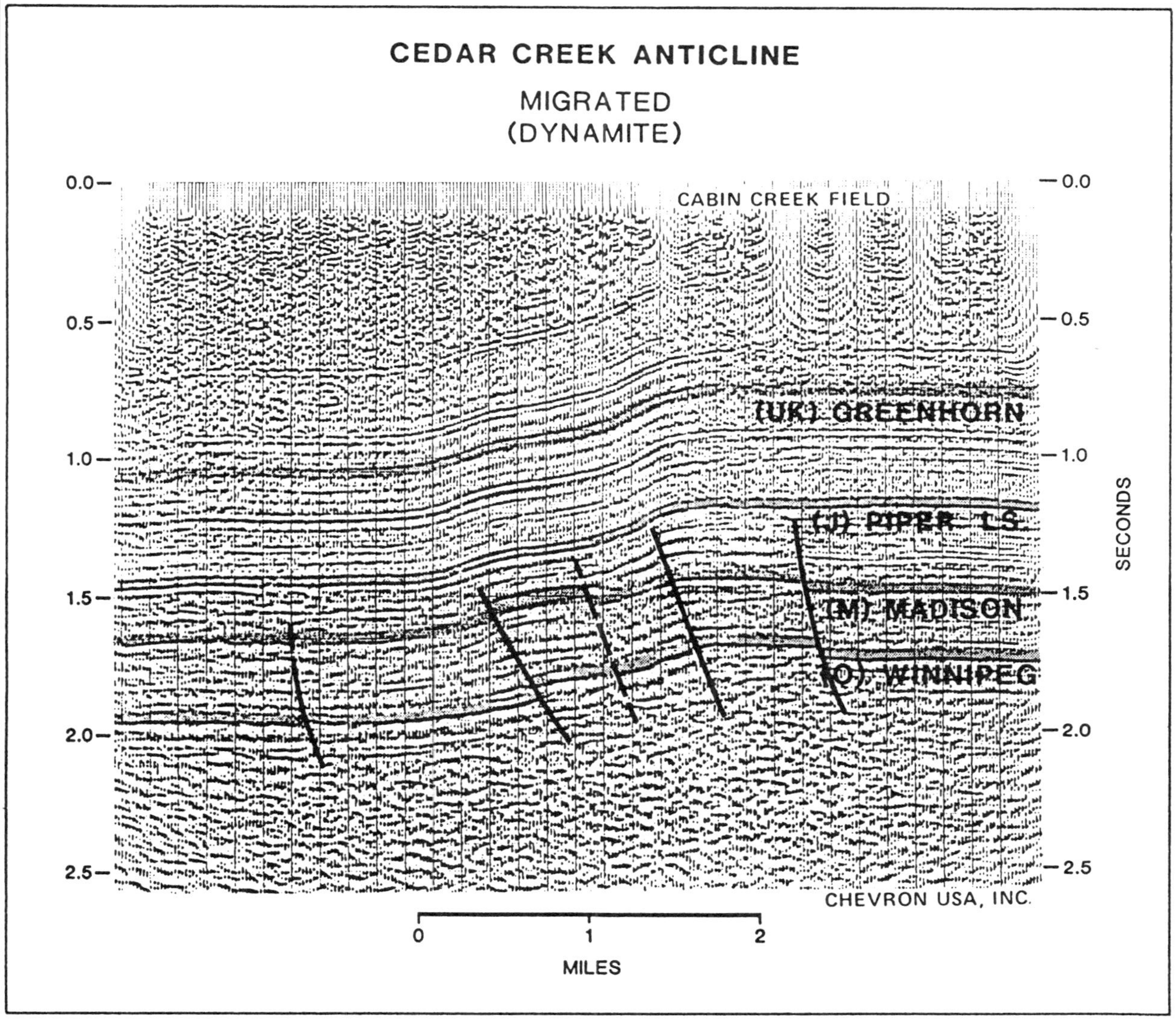

Seismic line across Cabin Creek field. Line of section is shown in Figure VII-5 here-after/ci-après

Fig. III-8 - Drape folds : Cedar Ck. (Montana USA)
(CLEMENT)

- *Plis de drapage : Cedar Creek (Montana, USA),*
(CLEMENT)

STEARNS
(1971)

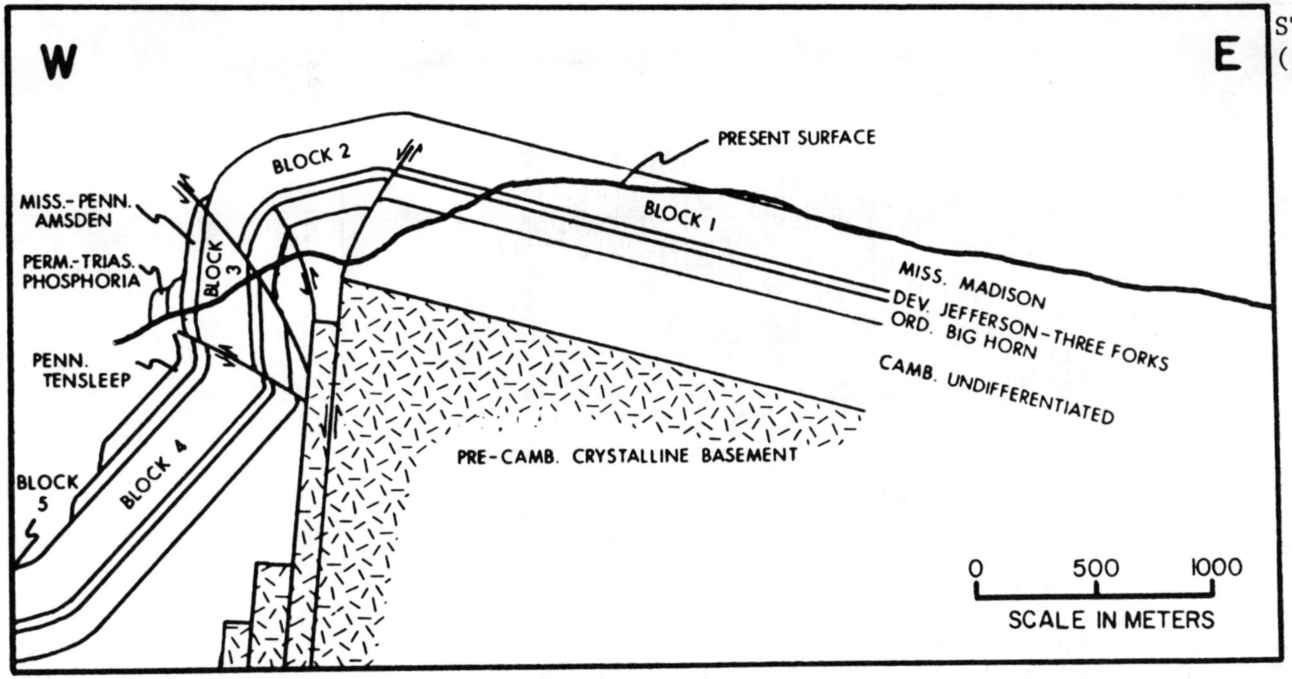

Cross section of Rattlesnake Mountain west of Cody, Wyoming. Stearns, 1978.

Fig. III-9 - Drape folds : Wyoming Province (USA)
(STEARNS)

- *Plis de drapage : Province de Wyoming (USA).*
(STEARNS)

FIGURES

IV. IDENTIFICATION CRITERIA FOR WRENCH FAULTS

CRITERES D'IDENTIFICATION DES FAILLES DE DECROCHEMENT

IV-C) FIGURES

IV-C) FIGURES

CROWELL (1962)

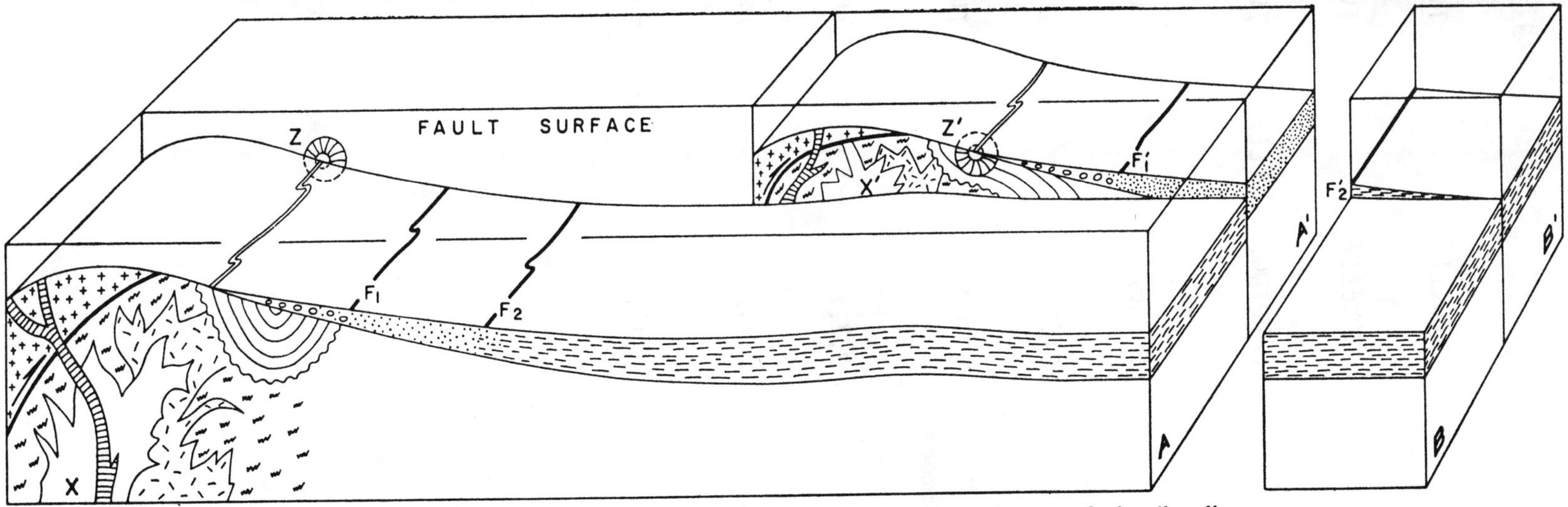

Regional trace slip and use of displaced geological "lines" to find strike slip

Regionally flat-lying sedimentary formation consisting of conglomerate, sandstone, and shale and lying unconformably on older rocks is displaced by pure right slip. Net slip indicated by displacement of several geologic "lines;" wedge-out line (Z–Z′), line of facies changes from conglomerate to sandstone (F_1–F_1') and from sandstone to shale (F_2–F_2'). Right slip results in trace slip primarily so that no separation at right. Cross section B–B′ displays no stratigraphic evidence of major fault and structure contours in this vicinity will cross fault without deflection. Dip separation and facies and thickness contrast across fault at A–A′.

Older terranes in vicinity of X and X′, schematically analogous to situation across San Andreas fault system in southern California; contain variety of rocks with same petrology, geologic history, and sequence of events recorded. Although there are numerous contacts with lines of intersection between them, the complexity of the geology and lack of data and outcrops interfere with use of these lines for finding slip. Separation of distinctive steeply dipping contacts may at times be interpreted as slip. Correlation under these circumstances will depend first upon comparison of rocks and recorded histories, and on regional knowledge that these rocks and histories are not present in intervening country, buried under other deposits, nor eroded away. With more data unique geological "lines" may be found so that slip can be established unequivocally.

Fig. IV-1 - Displaced geological lines (CROWELL)

Fig. IV-1 - Lignes géologiques décalées (CROWELL)

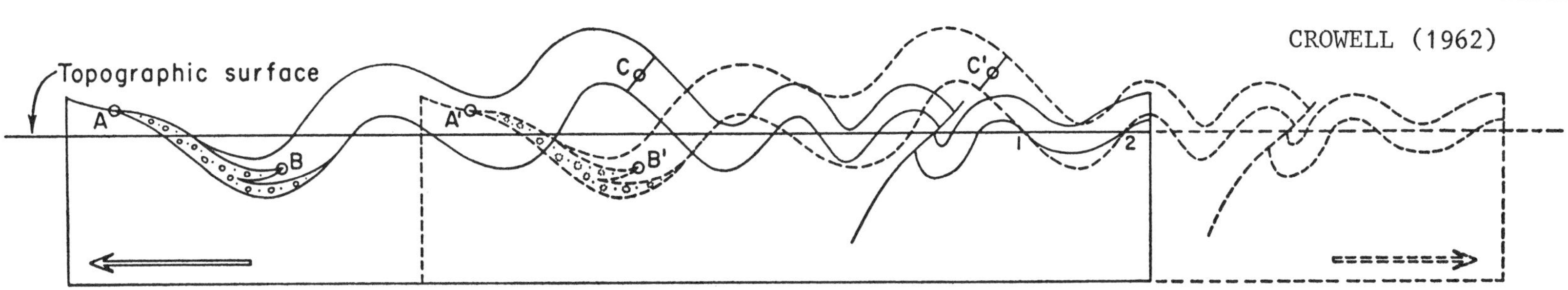

SECTION IN FAULT PLANE

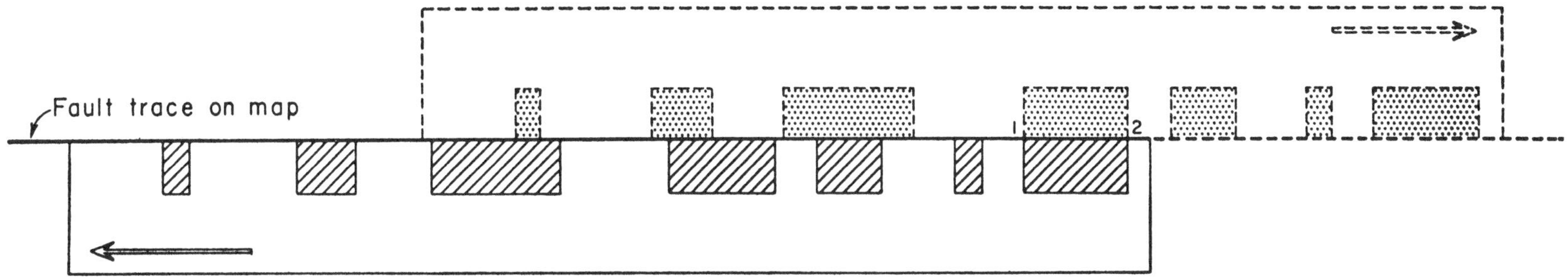

MAP ALONG FAULT

Regional trace slip and unsystematic separations resulting from major strike slip

At top is structure section in fault plane of hypothetical right-slip fault with vertical dip. Stratigraphic unit, folded and faulted before strike-slip faulting, displaced with right slip equal to A–A′, B–B′, and C–C′. Traces of contacts on the fault plane solid on near side of fault and dashed on far side. Slip defined where piercing points of three geological "lines" intersect fault plane. A and A′ are piercing points of wedge-out or basin-margin line of stratigraphic unit; B and B′ are piercing points of line of facies change; C and C′ are piercing points of midpoint of specific isopachus line.

Below is sketch map along fault. Note that separations are unsystematic although slip is constant. Some separations are to right and some to left; at two places (1, 2) contacts are traced across fault without apparent displacement. Only by intensive study of unit to find wedge-out, facies-change, and isopachus lines can slip be determined.

Fig. IV-2 - Separations
(CROWELL)

Fig. IV-2 - Séparations
(CROWELL)

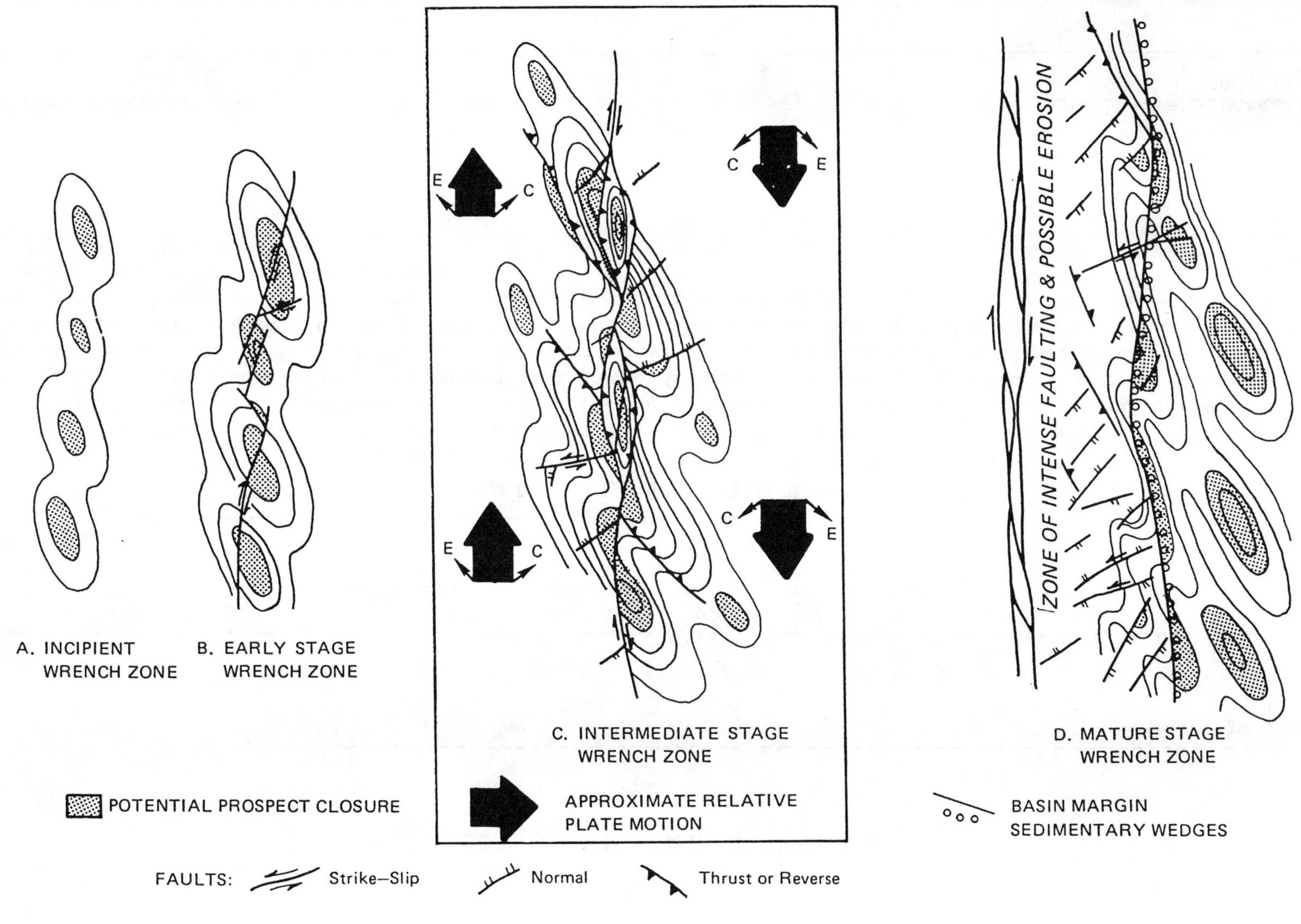

Schematic diagram of structural assemblage associated with major wrench fault and stages in evolution that cause changes in hydrocarbon trapping; modeled after California examples (discussed in Harding, 1974). Arrows C, E represent compressive and extensional vectors, respectively, that arise from coupling motion between adjacent plates.

HARDING & LOWELL (1979)

Fig. IV-3 - Wrench-corridor with folds and faults
(HARDING, LOWELL)

- *Couloir de cisaillement, avec plis et failles*
(HARDING, LOWELL)

LINEAMENTS: THEIR ROLE IN TECTONICS OF CENTRAL ROCKY MOUNTAINS

STONE (1974)

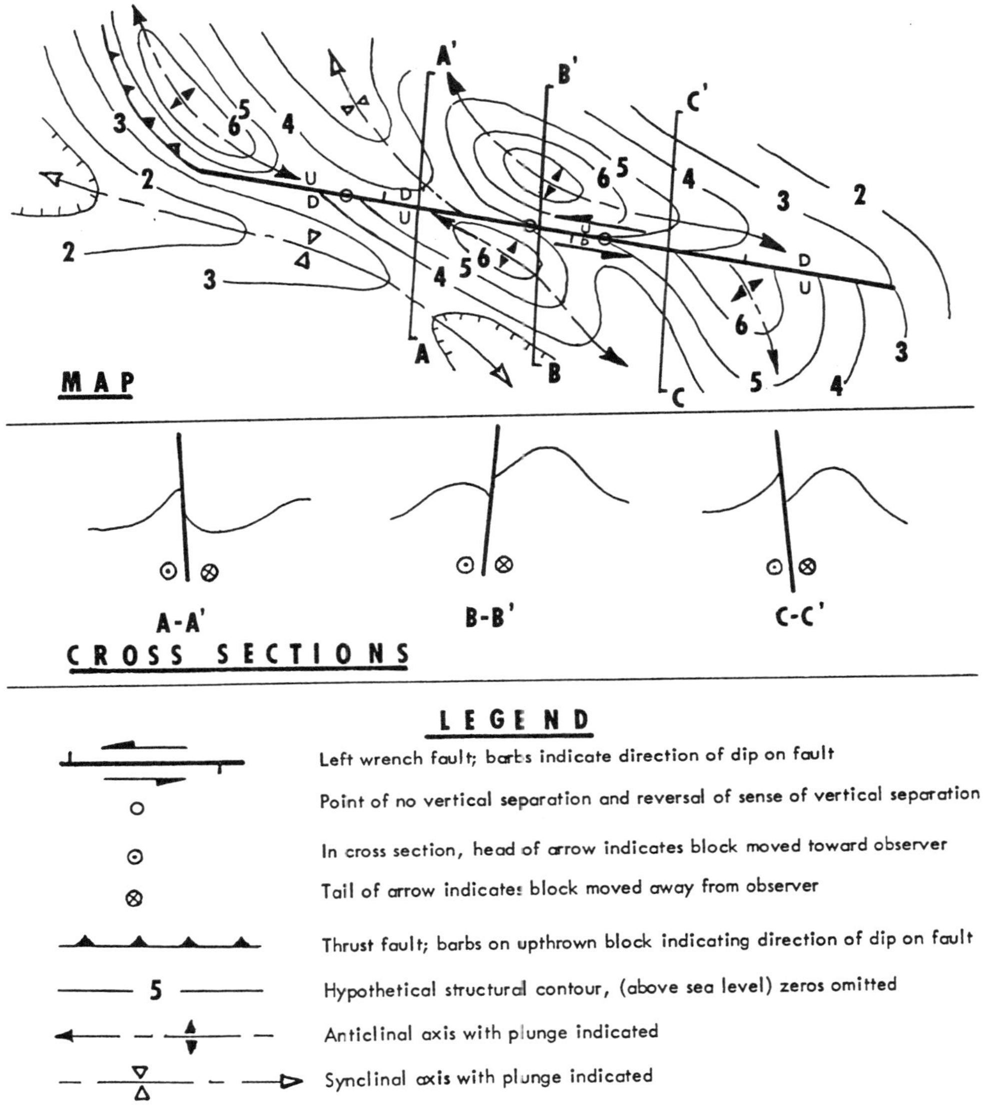

Diagrammatic structural contour map and vertical cross sections illustrating "reverse drag" associated with wrench faulting. (see legend)

Fig. IV-4 - Drag-effect (STONE)

Fig. IV-4 - Effet d'entraînement (STONE)

STONE
(1974)

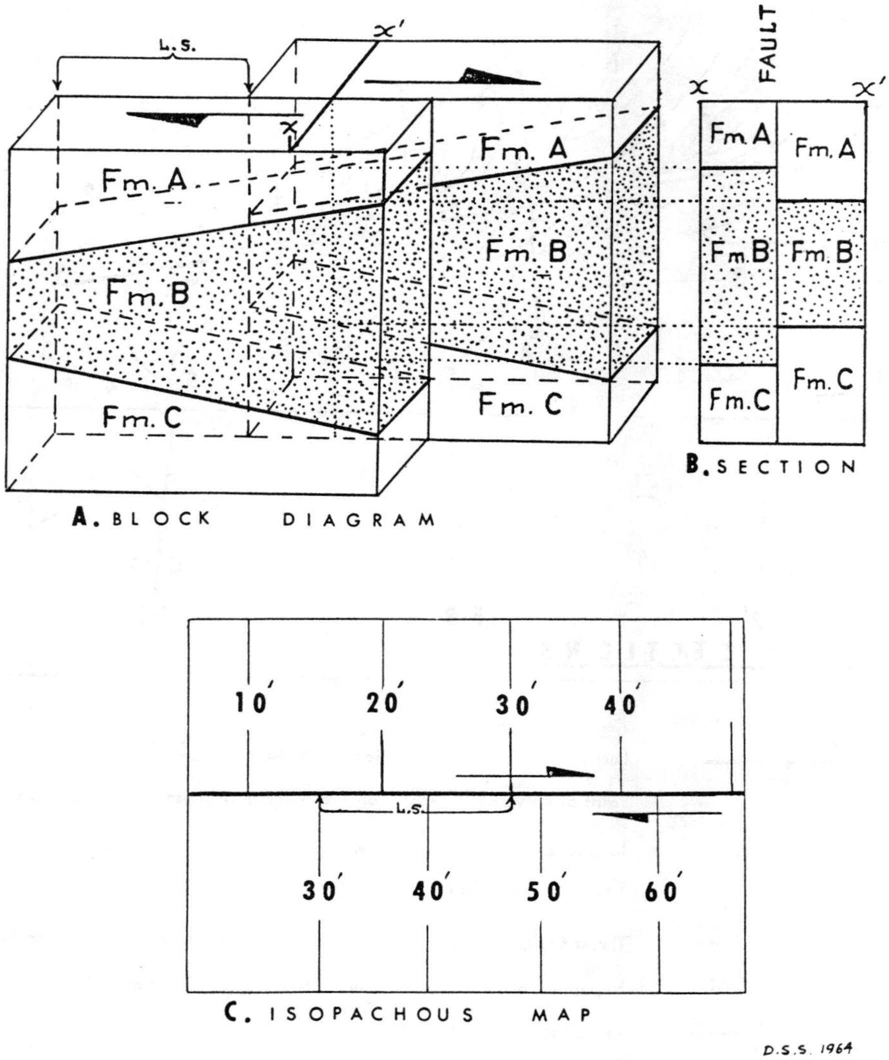

(A.) Block diagram, (B.) vertical section, and (C.) isopachous map, along a wrench fault cutting a formation, "B", which changes thickness along the fault. The resulting cross sectional view illustrates why this condition is so often interpreted as due to reversals of vertical movement through time (i.e., "bouncing blocks"). L.S. = lateral slip.

Fig. IV-5 - Thickness shift
(STONE)

- *Décalage des valeurs d'épaisseur*
(STONE)

(KINGMA)

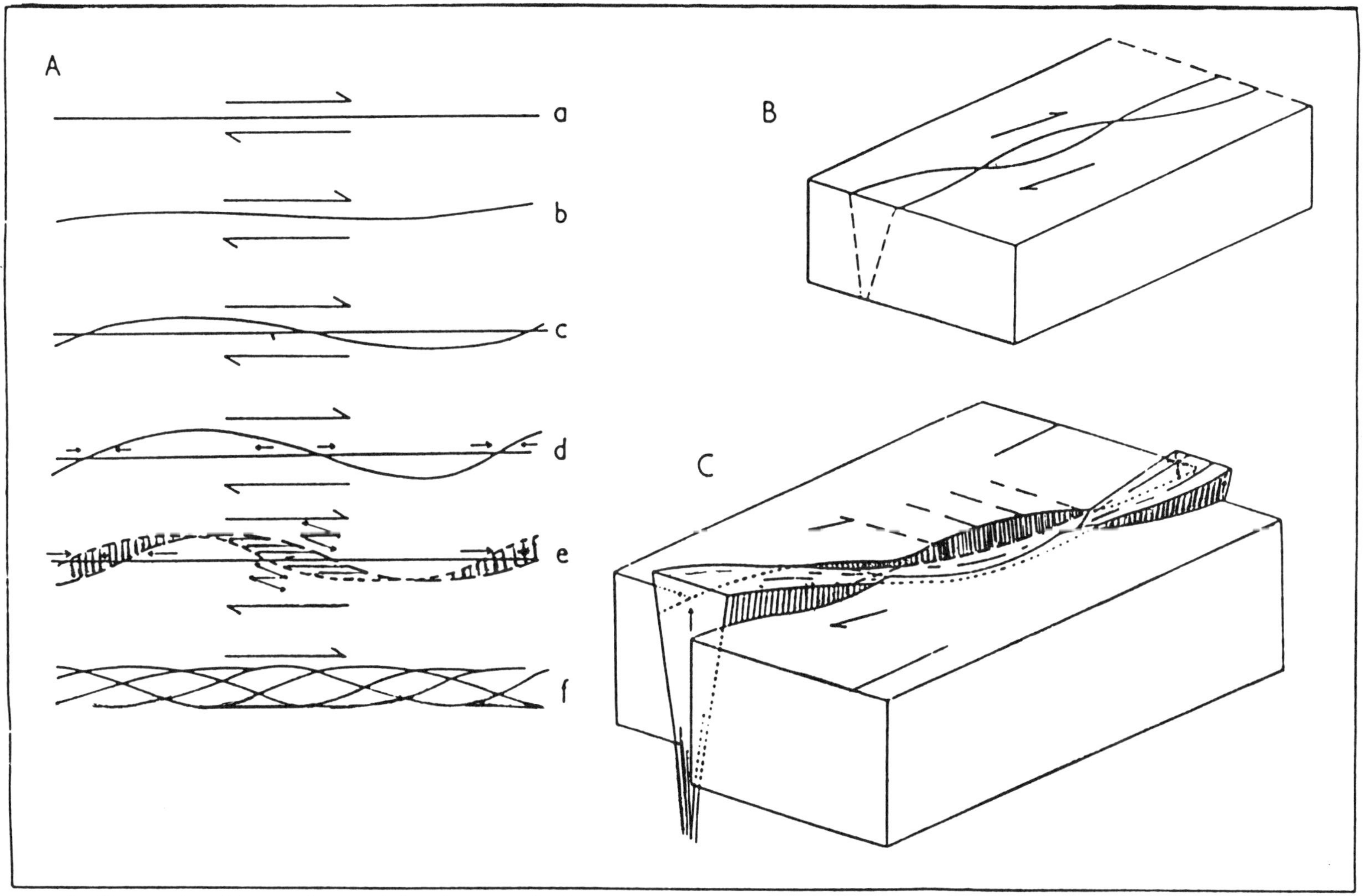

.—A. Stages in deformation and in change of direction along part of a dextral transcurrent fault; — denotes tensional and sinking areas; + compressional and rising areas. B. Dextral transcurrent fault-zone before movement. C. same fault-zone after movement has taken place; development of horsts and graben.

Fig. IV-6 - Stages of deformation of a wrench-zone and associated horsts and graben

- *Stades de déformation d'une zone de cisaillement et ses horsts et graben* (handwritten correction: "et" struck, "de" inserted)

EXISTENCE D'UN RÉSEAU DE FAILLES DE DÉCROCHEMENT...

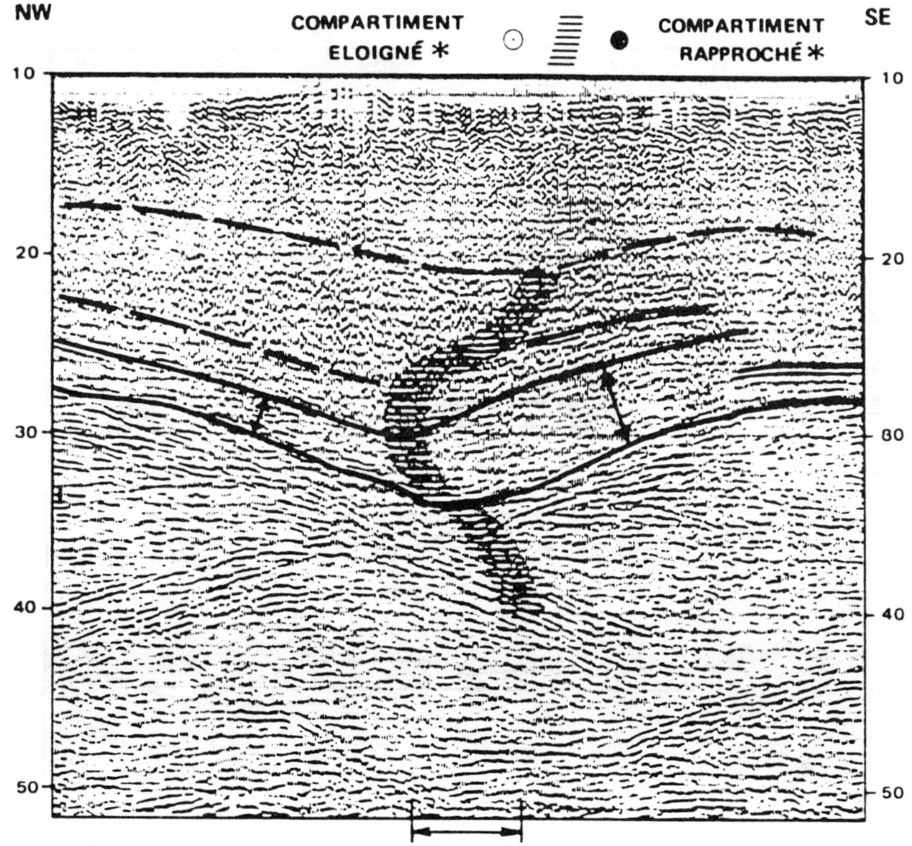

Encoignure sismique observée souvent à l'aplomb des décrochements

(A noter les différences d'épaisseur de part et d'autre)

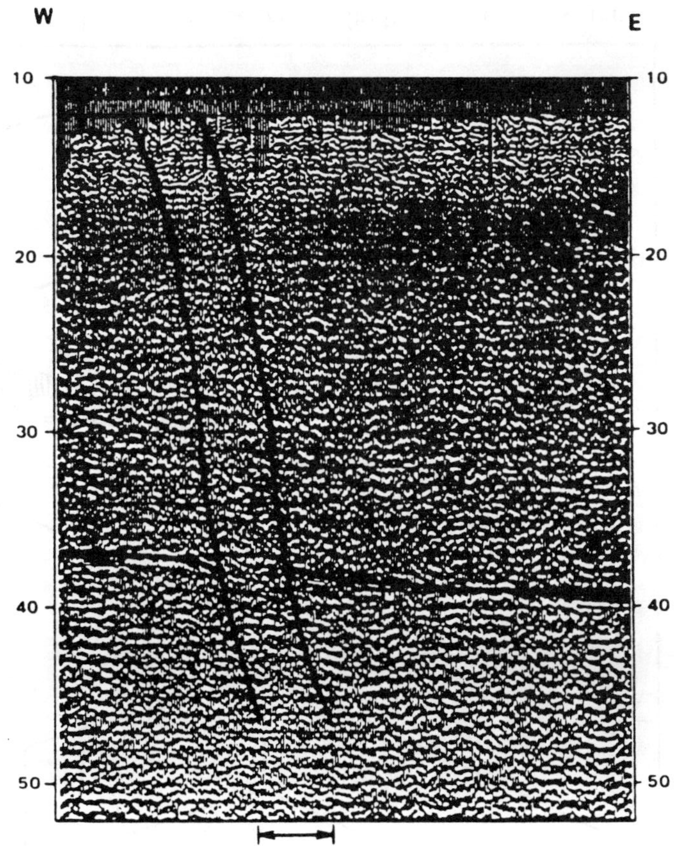

Zone brouillée

* *selon données de terrain*

Fig. IV-7 - Seismic discontinuity (G.Z.)

Fig. IV-7 - Discontinuité sismique (G.Z.)

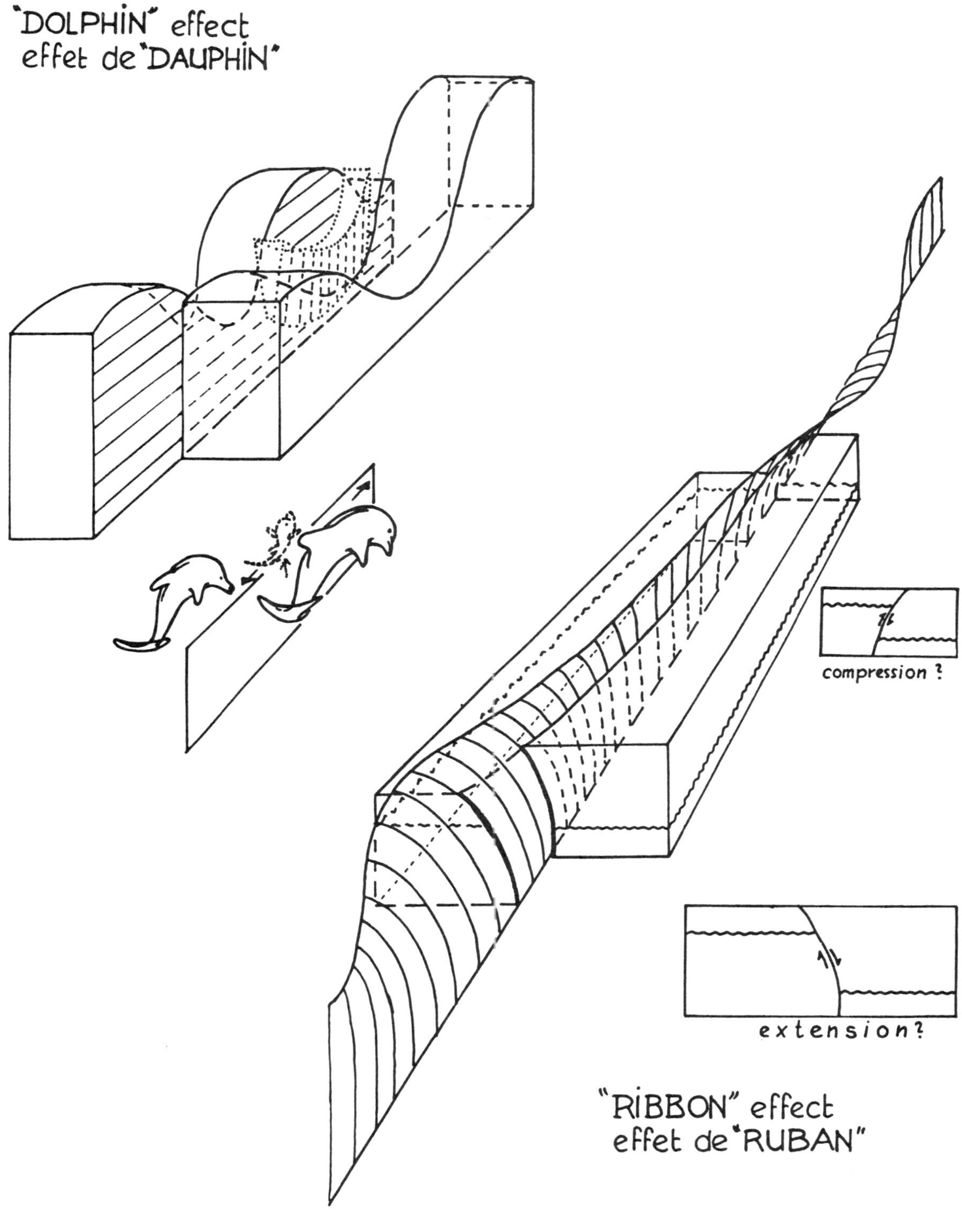

Fig. IV-8 "Dolphin effect"
"Ribbon effect"

Fig. IV-8 "Effet de Dauphin"
"Effet de Ruban"

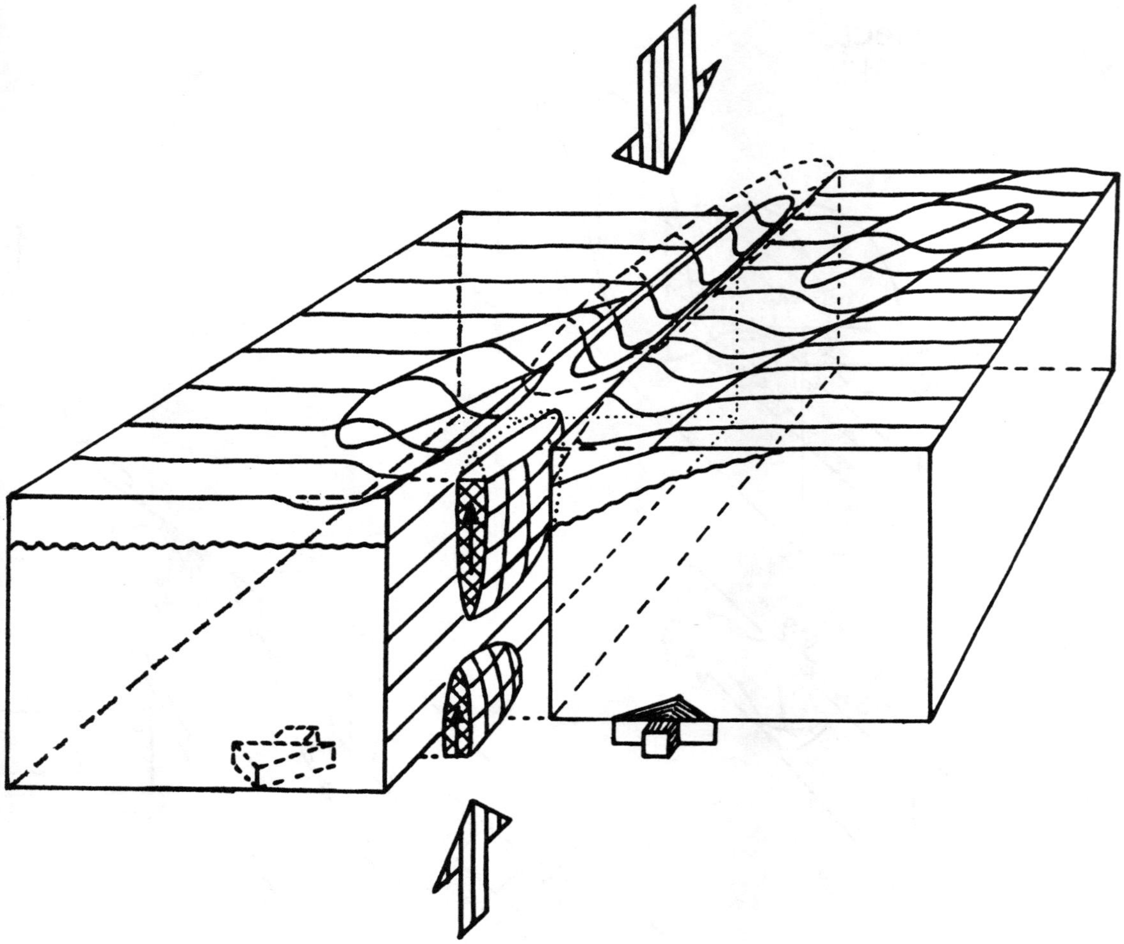

Fig. IV-9 Verticalness of wrench-corridors

Fig. IV-9 Verticalité des zones de cisaillement

FIGURES

V. AMPLITUDE OF MOVEMENTS

AMPLITUDE DES MOUVEMENTS

V-B) FIGURES

V-B) FIGURES

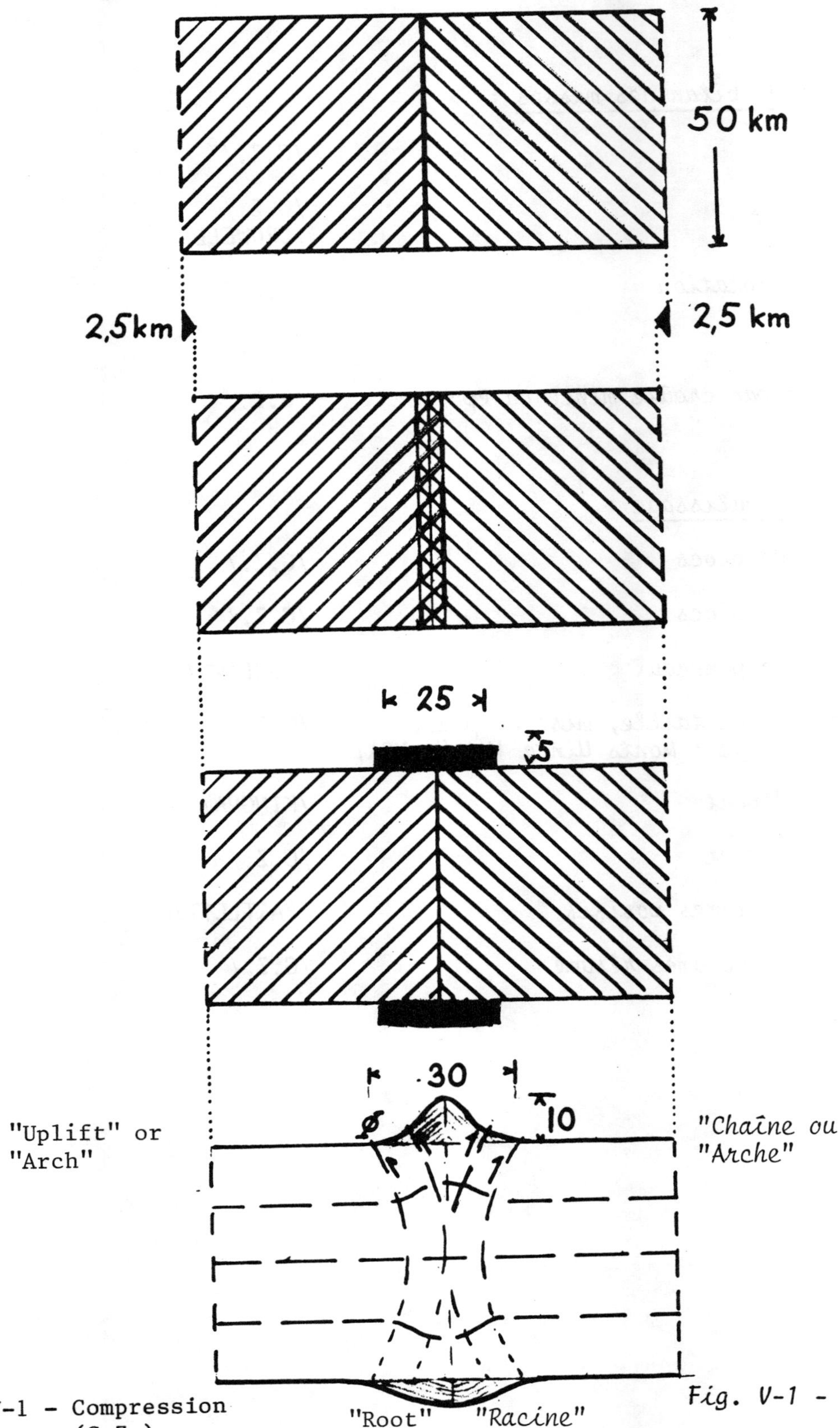

Fig. V-1 - Compression (G.Z.)

Fig. V-1 - Compression (G.Z.)

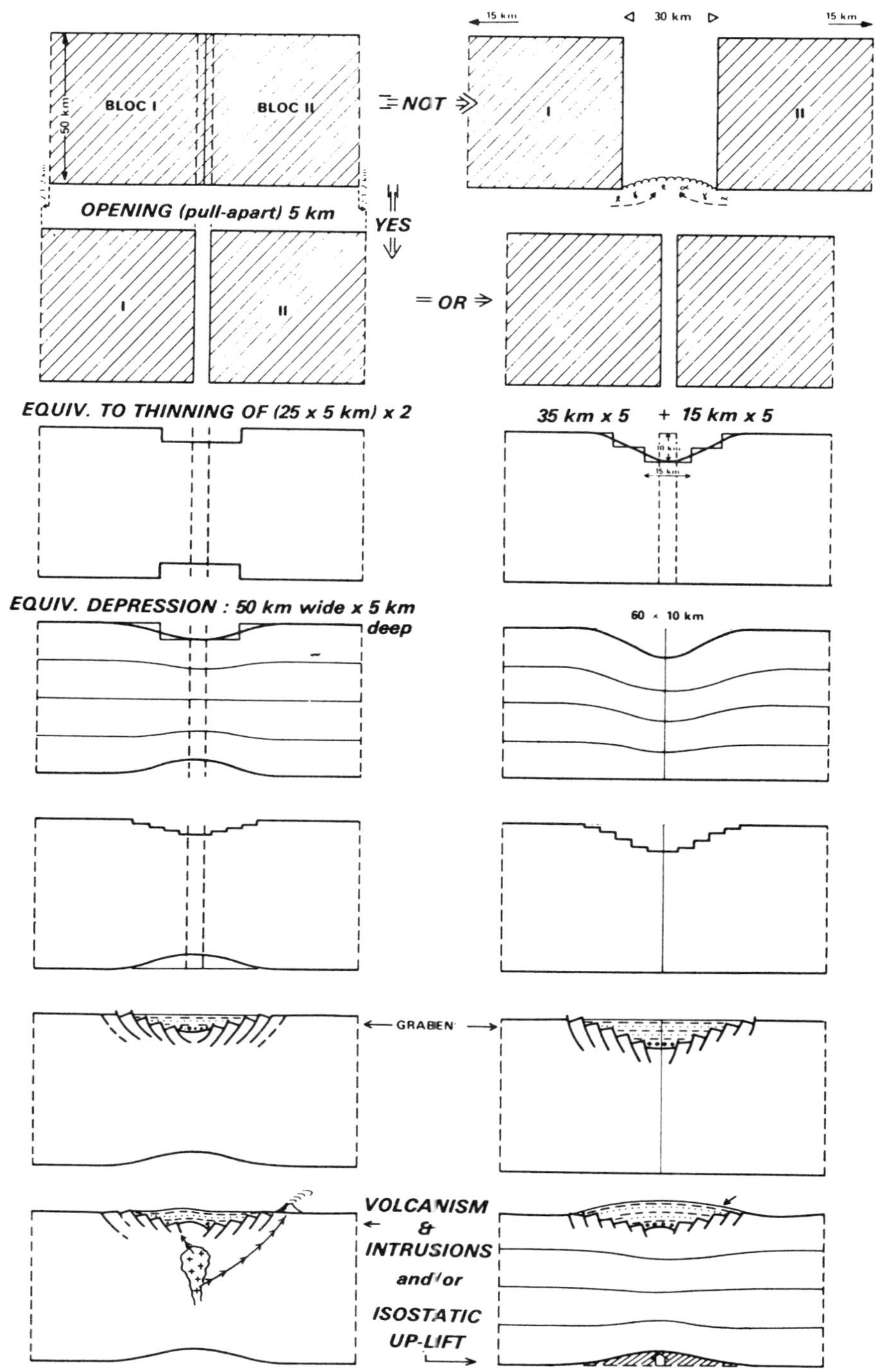

Fig. V-2 - Extension/streching
(G.Z.)

Fig. V-2 - Etirement
(G.Z.)

CROWELL (1974)

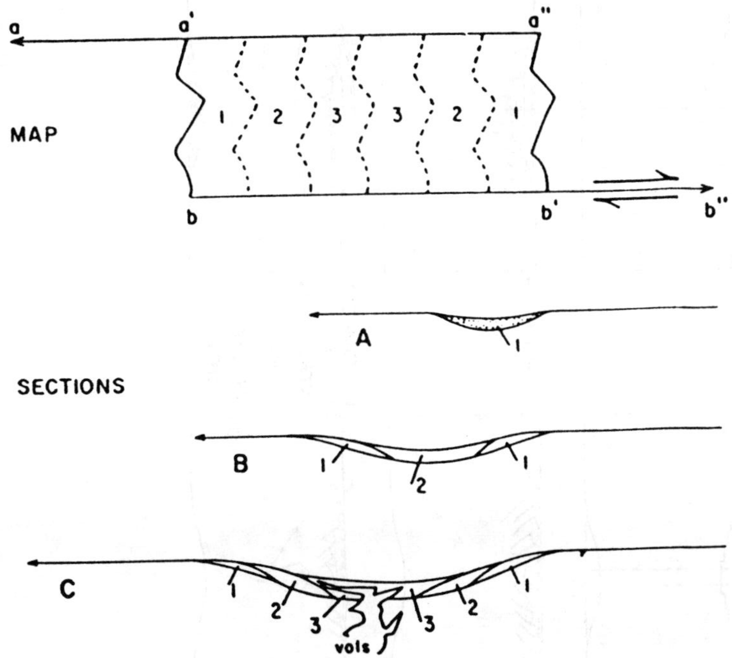

—Sketch map and sections of pull-apart basin;

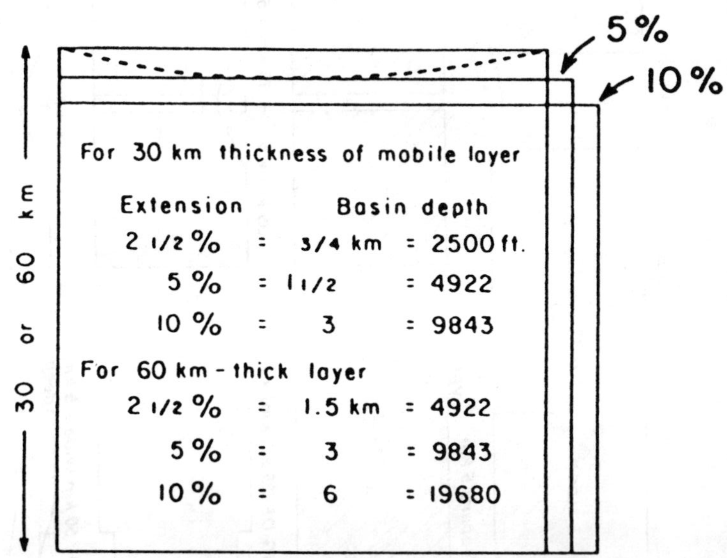

—Diagram showing crustal cross section, 30 or 60 km thick, with 5 or 10% extension

Fig. V-3 - Extension/streching (CROWELL)

Fig. V-3 - Etirement (CROWELL)

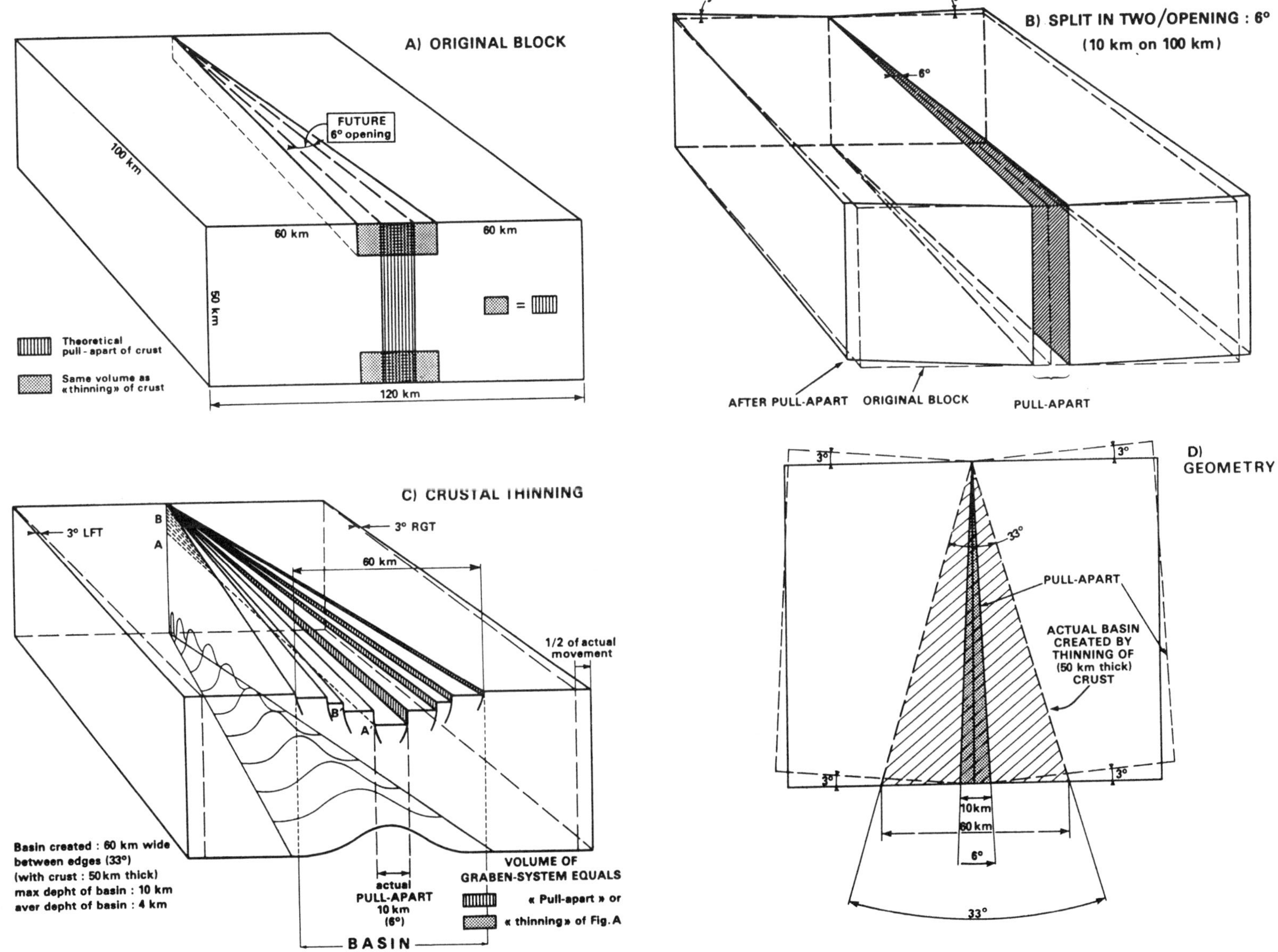

Fig. V-4 - Opening/rotation (G.Z.)

Fig. V-4 - Ouverture-rotation (G.Z.)

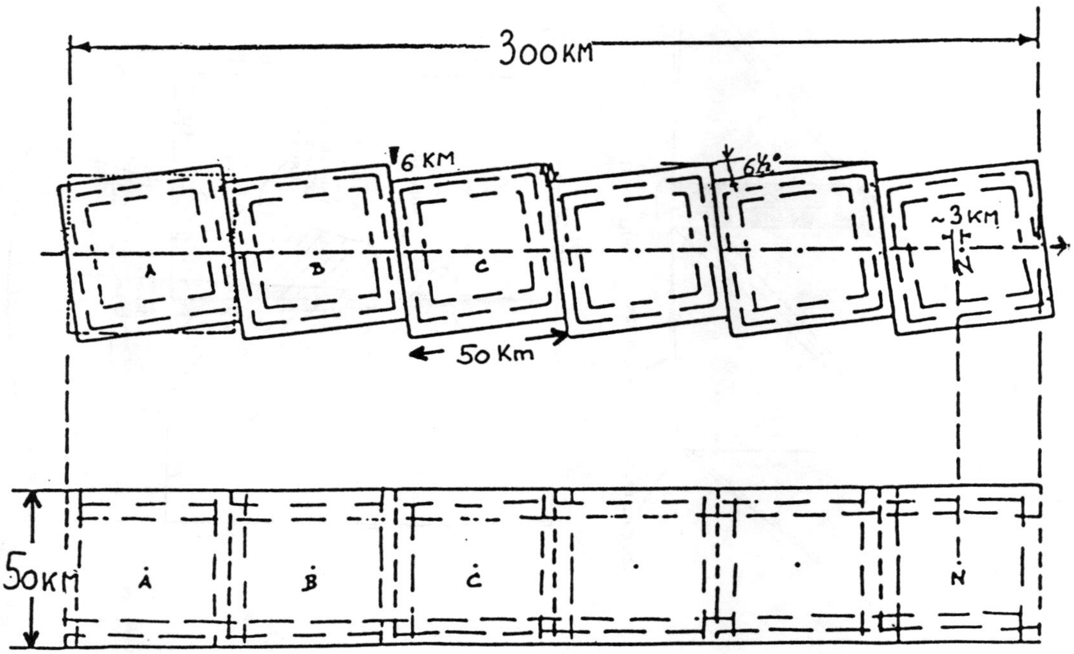

A stretching of 1 % (here : 3 km/300 km) may produce a series of tilted blocks, 50 km wide ea., in a crust 50 km thick. The maximum depth of the half-graben will be of 6 km, the dip of the basin-(graben) bottom will reach 6 1/2°.

Par l'étirement d'1 % (ici : 3 km/300 km) d'un segment crustal épais de 50 km, on pourra obtenir une suite de blocs basculés de 50 km de largeur ; la profondeur de chaque demi-graben sera alors de 6 km, tandis que le pendage du fond atteindra 6,5°.

Fig. V-5 - Tilting (G.Z.)

Fig. V-5 - Basculement (G.Z.)

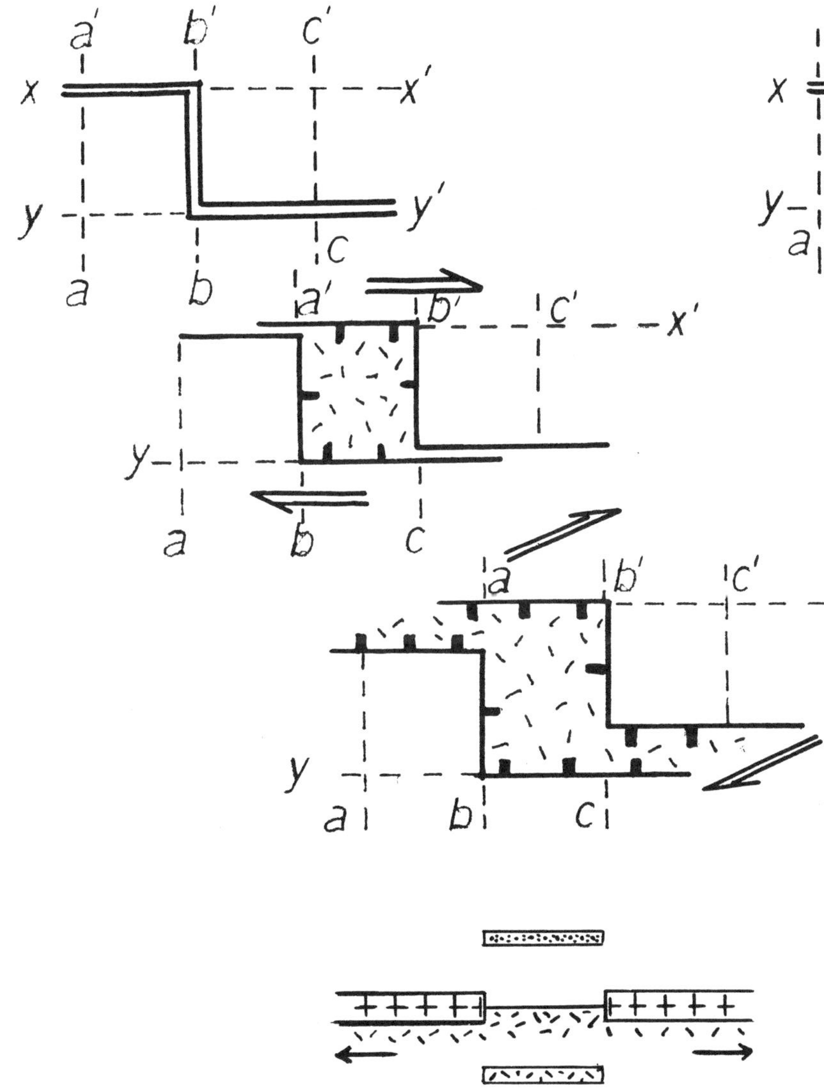

Fig. V-6 - Extension with thick/thin crust
(G.Z.)

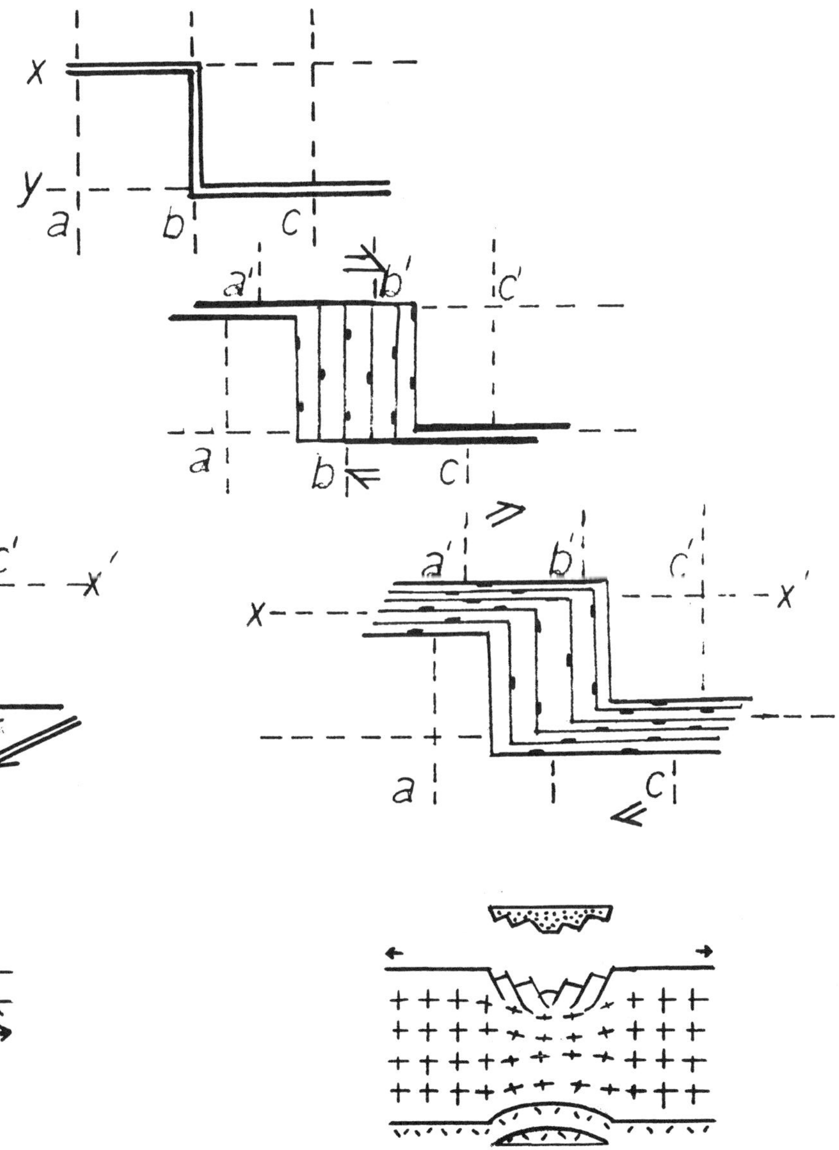

Fig. V-6 - *Extension par croûte mince ou épaisse*
(G.Z.)

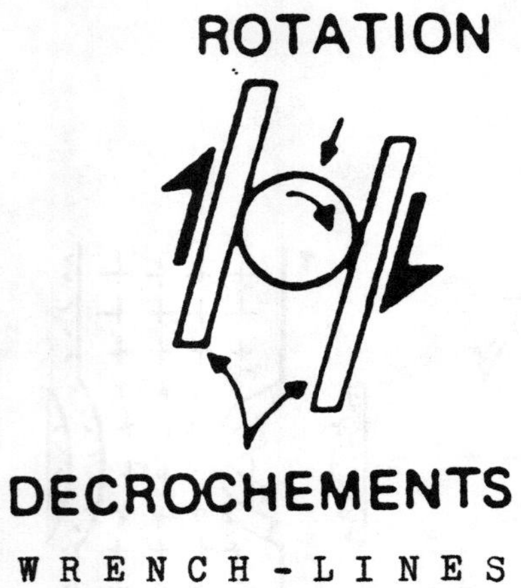

Fig. V-7 - Rotation of blocks
(G.Z. 1975)

Fig. V-7 - Rotation de blocs
(G.Z. 1975)

" Rack-and-pinion " gear effect ...

Effet de " direction à crémaillère "...

Fig. V-8 - Uplifting of blocks
(G.Z. 1985)

Fig. V-8 - Ejection de blocs
(G.Z. 1985)

BLOCK DIAGRAM

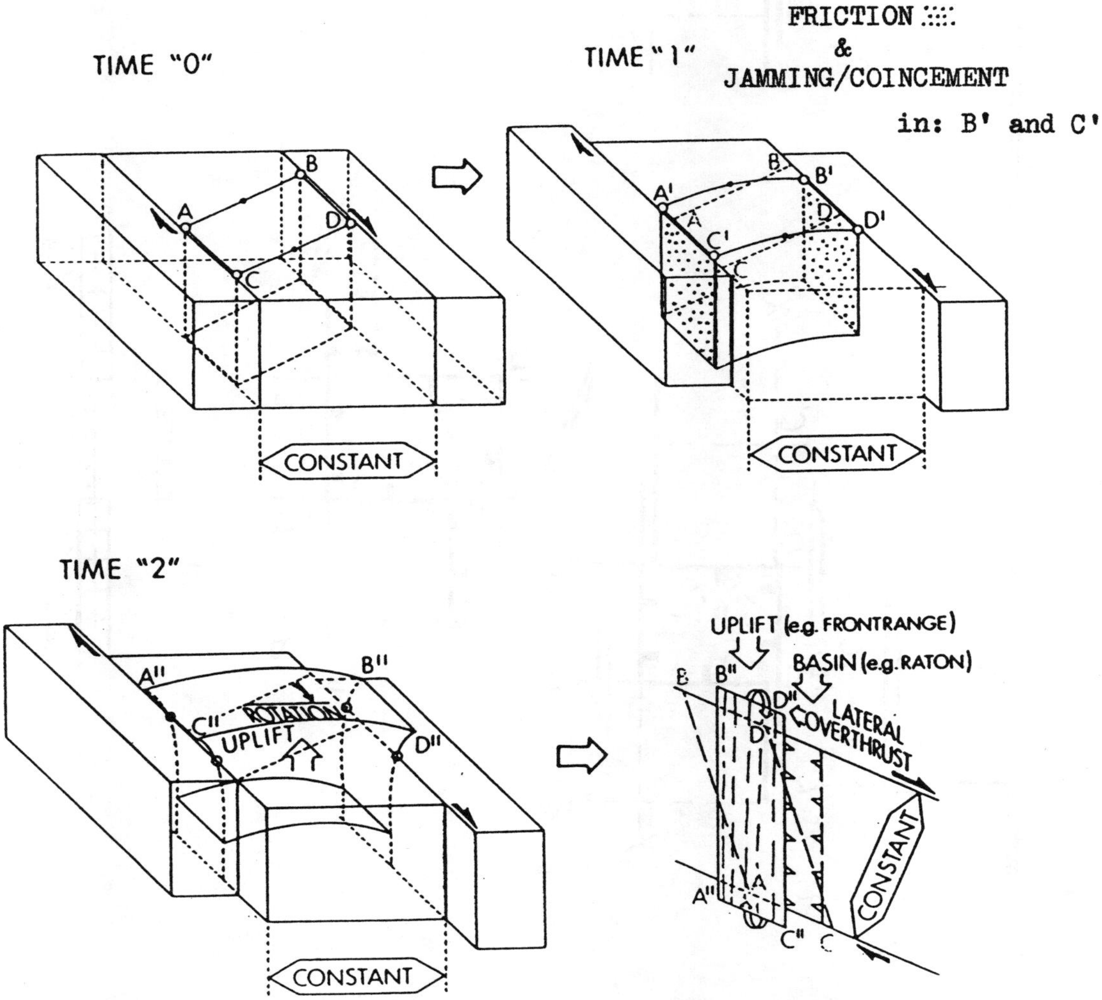

"RUHLAND's RULER"-RULE

(LA RÈGLE de la RÈGLETTE de RUHLAND) } "R-R-R experience"

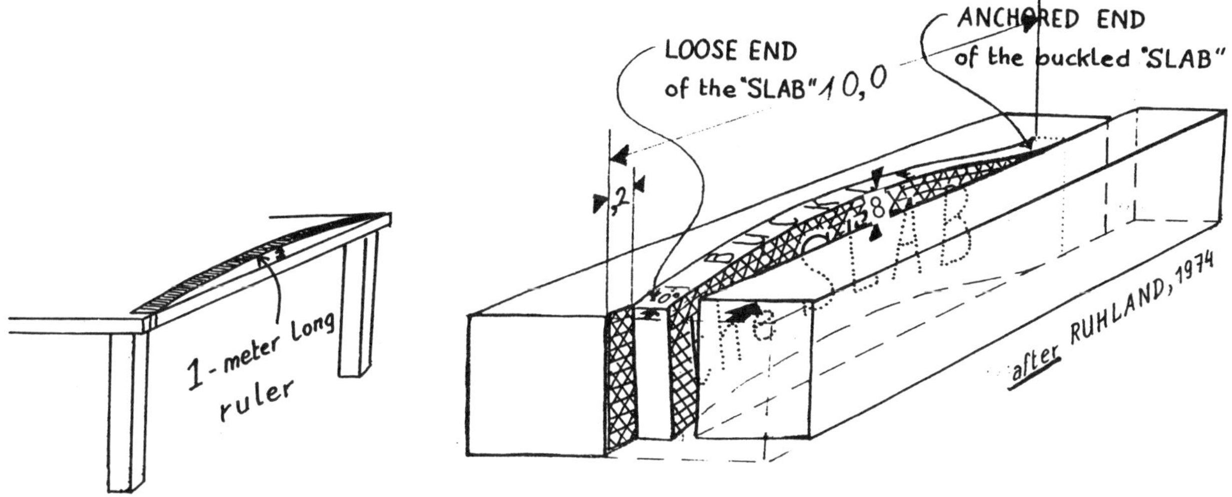

E.g. — a 10 km long « slab », taken into a « wrench-corridor »

— compressed (shifted) horizontally on 0,2 km (on the near end) [very difficult to see in seismic/subsurface]

— buckles up by 0,8 km at the highest of the vault [very easily recognized by any method]

« Check with a 100 cm ruler on the edge of your desk »

after RUHLAND, completed & modified (G.Z., 1987)

d'après RUHLAND, complété et modifié (G.Z., 1987)

Fig. V-9 - Vaulting of "slabs"

Fig. V-9 - Voussure de panneaux

See also Fig. VI-3 - *Cf. Fig. VI-3*

Fig. V-10 - Large-scale uplift related to a mega-shear :
Uinta Range (Utah, USA)

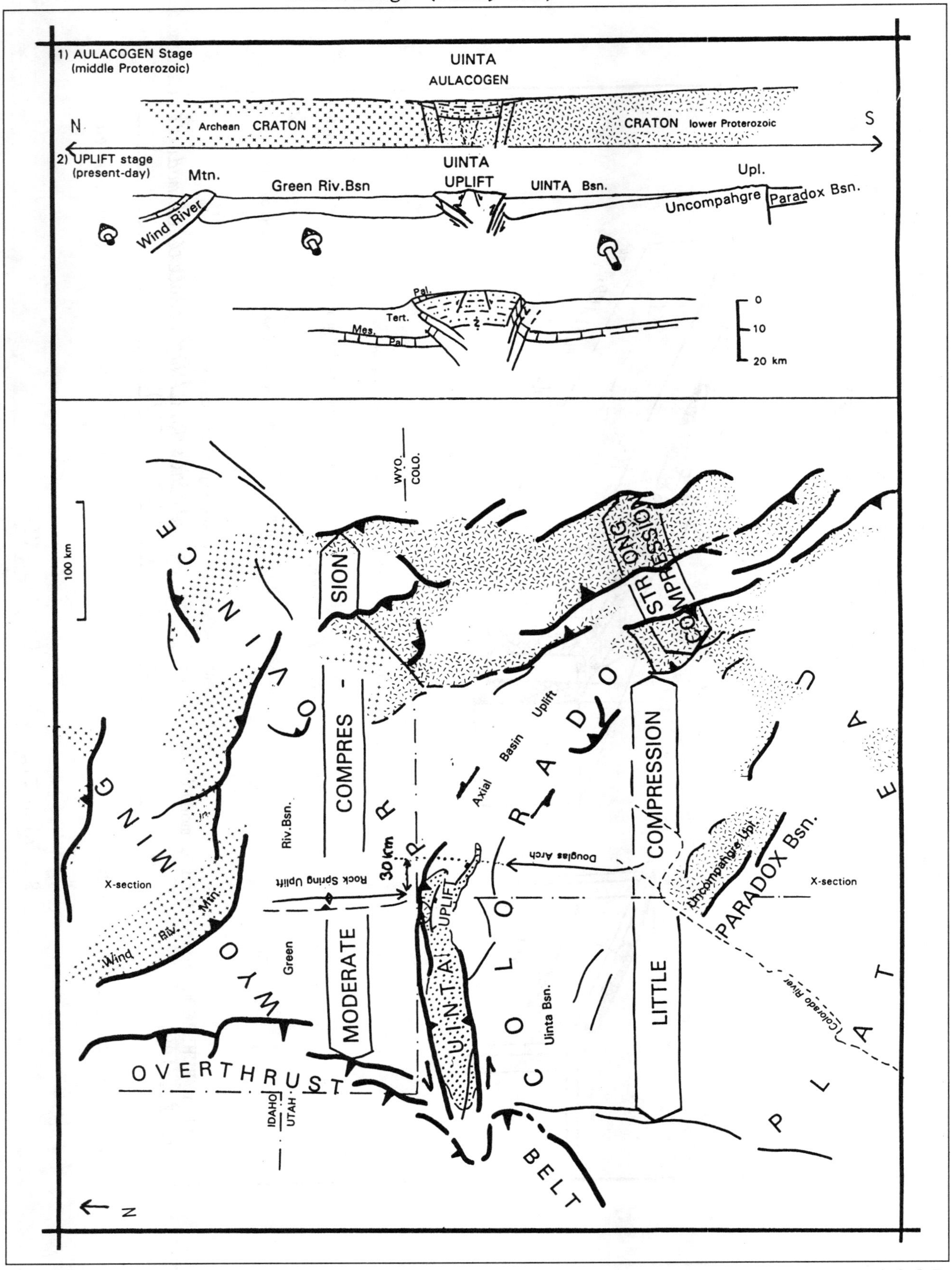

A.A.P.G. - NICE - 1988

Short course n° 2 G.Z.

Fig. V-10 - Un "uplift" de taille, mise en place par coulissage :
Monts Uinta (Utah, USA)

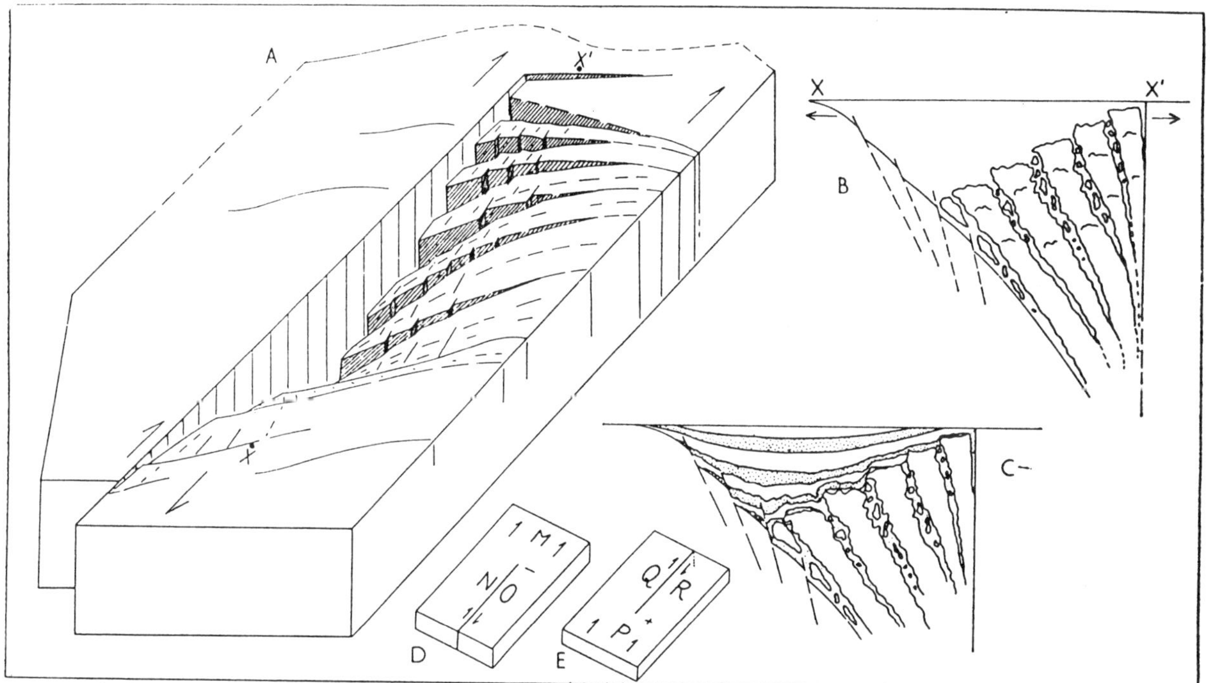

—A. Dextral transcurrent fault petering out, area to right under tension with consequent collapse. B. Section through collapsed area in A, drawn between points X–X', showing tectonic deformation and breaking up of local "undermass". (Syn-tectonic sediment cover taken off.) C. Syn-tectonic sedimentary cover restored over broken "undermass". D. Dextral transcurrent fault petering out; — denotes tensional area. E. Dextral transcurrent fault beginning; + denotes rising area.

Fig. V-11 - Perte de déport
(KINGMA)

Fig. V-11 - Decreasing displacement
(KINGMA)

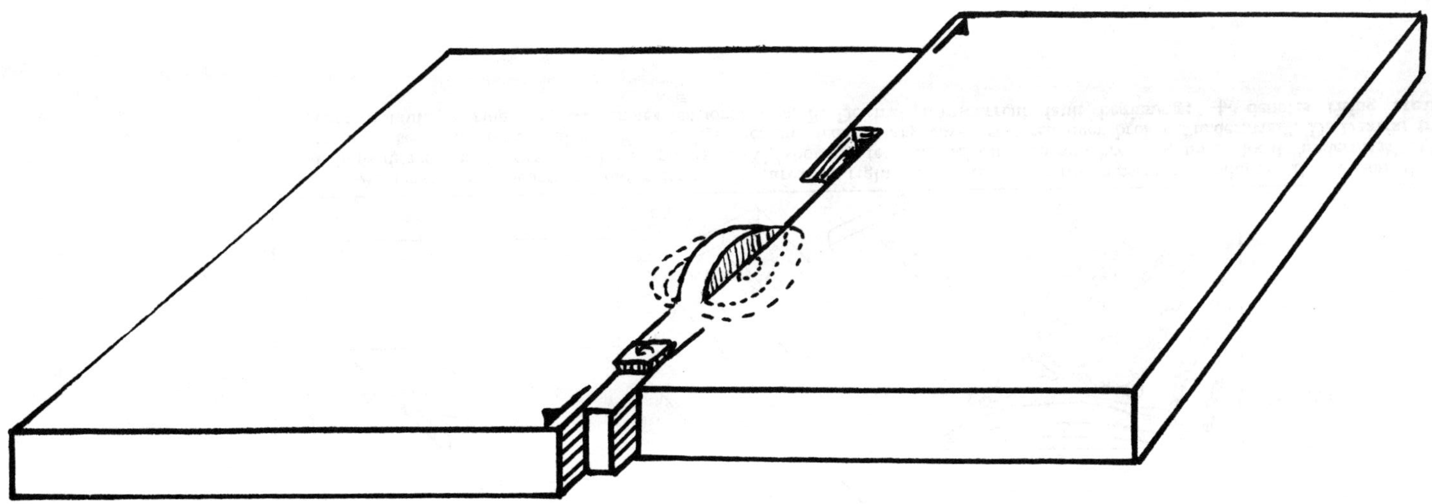

Fig. V-12 - Constant displacement
(G.Z.)

Fig. V-12 - Déport constant
(G.Z.)

From/d'après: BARTLETT et al. 1981

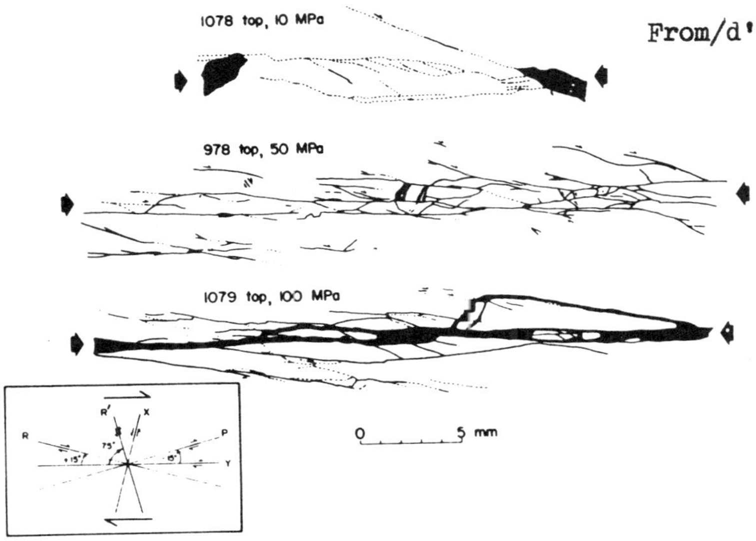

. Right-lateral, shear-zone fracture maps, top sections, of 0.5-cm thick layers of Indiana limestone deformed at 10- (Sp. 1078), 50- (Sp. 978), and 100-MPa (Sp. 1079) confining pressure, and taken to 0.46, 0.54, and 0.56-cm shear displacement, respectively. Solid lines indicate fractures along which there is separation; dashed lines are traces of gouge. Solid black areas indicate missing material (lost in thin section preparation). Bold arrows indicate orientation and location of sawcut in underlying forcing block. Boxed insert illustrates orientations and sense of shear for *R*-, *R'*-, *P*-, *X*-, and *Y*-shears in a right-lateral shear zone.

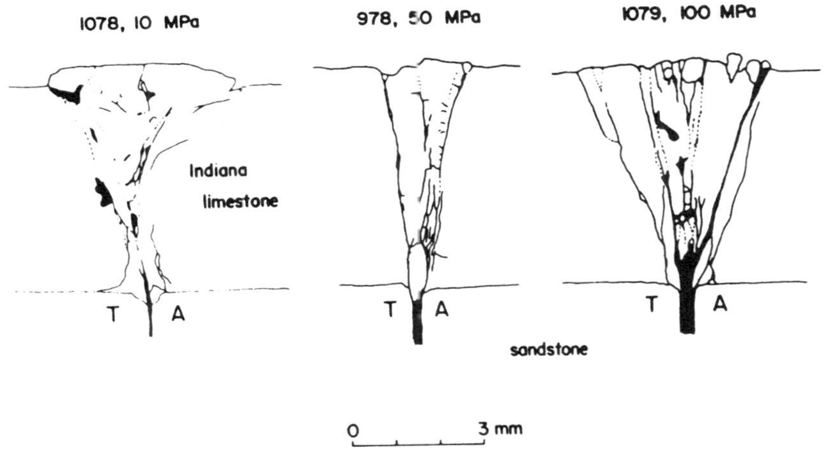

Cross-section, shear-zone fracture maps of 0.5 cm of same specimens shown **above.** Sense of shear indicated by *T* (towards) and *A* (away). Line character same as in **top figure.**

Fig. V-13 - Positive mass-balance (BARTLETT)

Fig. V-13 - Bilan des volumes positif (BARTLETT)

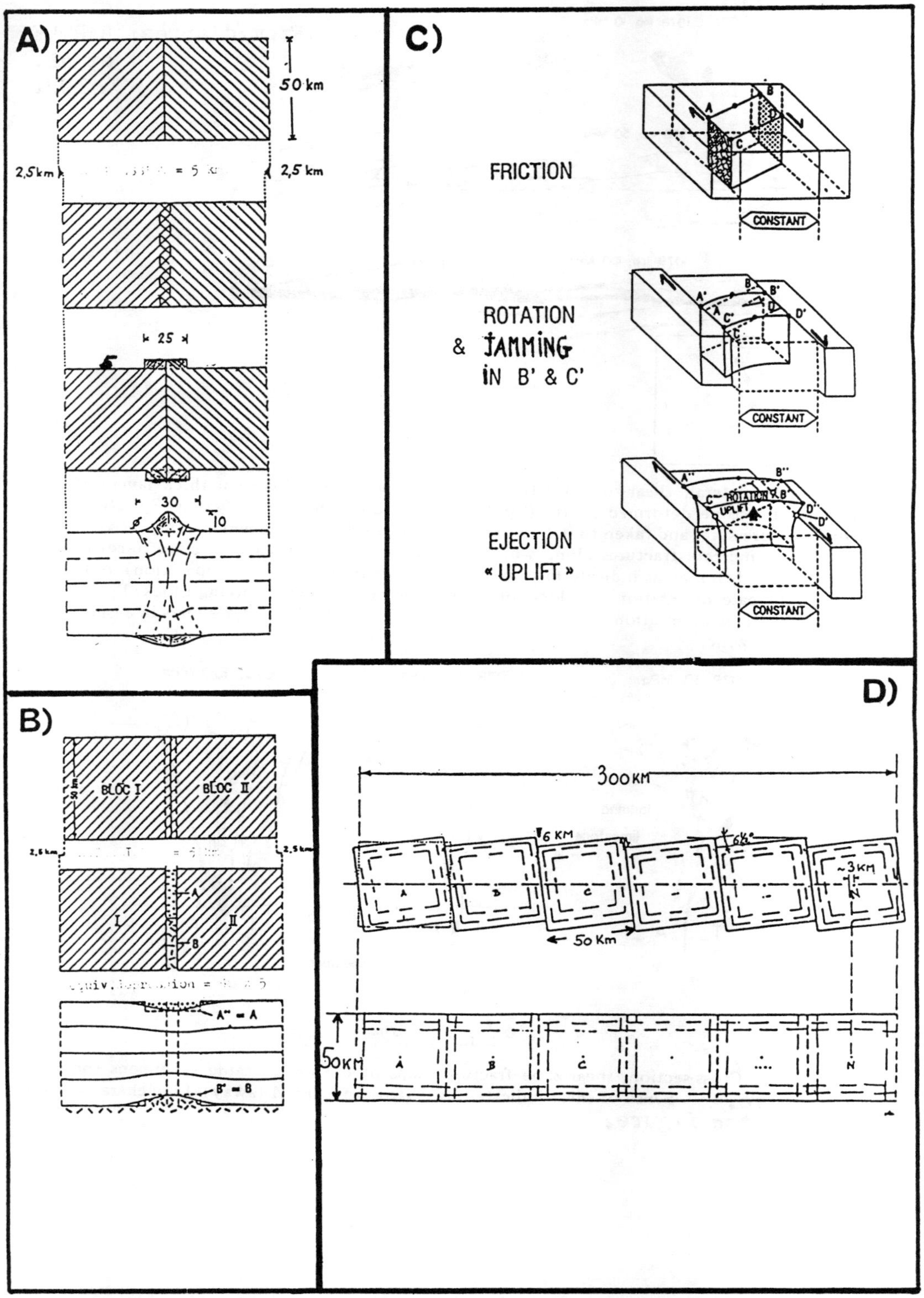

Fig. V-14 - Mass balance diagrams-recapitulation (G.Z.)

Fig. V.14 - Balance des volumes-résumé (G.Z.)

FIGURES

VI. ROLE OF WRENCHING IN THE CONTINENTAL ARCHITECTURE

ROLE ET INFLUENCE DES CISAILLEMENTS SUR L'ARCHITECTURE CONTINENTALE

VI-B) FIGURES

. The "Ancestral Rockies" system, late Paleozoic (W. USA)	/G.Z./	VI-19
	/U.S.G.S./	VI-19/B
"Ancestral" vs. "Laramide" Rockies	/G.Z./	VI-19/C-D-E
. Britanny (W. France)		VI-20
. The Central Australian Trough (Fitzroy - Amadeus)	/G.Z./	VI-21
. Western Indonesia - Malaysia	/G.Z./	VI-22
. Eastern France (Burgundy, Jura, Rhine Rhône Areas)	/RUHLAND, GELARD, BERGERAT,G.Z./	VI-23

4) <u>Consequences - "Side-effects"</u>

. Diapirism - halokinesis (Franklinian furrow, Arctic Isl.)	/FOX/	VI-24
. Gravity image (U.S.A.)	/G.Z./	VI-25

VI-B) FIGURES

. Le système des "Rocheuses Ancestrales" (Paléozoïque supérieur) (W.USA) Rocheuses "Ancestrales" et "Laramiennes"	/G.Z./ /U.S.G.S./	VI-19 VI-19/B VI-19/C-D-E
. Bretagne (France)		VI-20
. L'effondrement central d'Australie	/G.Z./	VI-21
. Indonésie - Malaisie	/G.Z./	VI-22
. L'Est de la France : Bourgogne, Jura Région Rhin-Rhône	/RUHLAND, GELARD, BERGERAT, G.Z./	VI-23

4) <u>Conséquences - Effets secondaires</u> :

. Diapirisme - Halocinèse (Sillon Franklinien, Canada)	/FOX/	VI-24
. Image gravimétrique (U.S.A.)	/G.Z./	VI-25

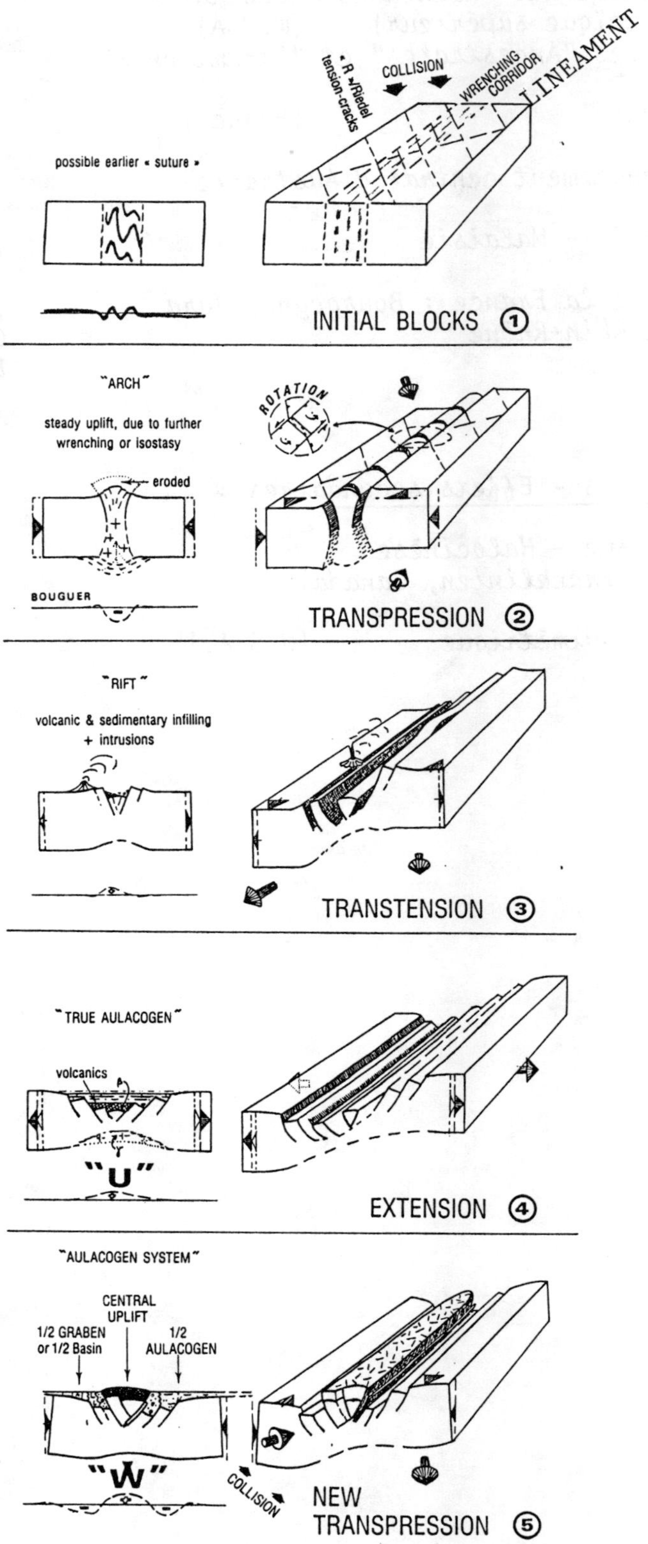

Fig. VI-1 - Lineaments - Arches - Rifts - Aulacogens - Uplifts (Evolutionary scheme of inter-related tectonic elements with wrench-components) (G. Z. 1985.)

Fig. VI-1 - Linéaments - Arches - Rifts - Aulacogènes - "Uplifts" (Schéma évolutif d'éléments tectoniques apparentés, avec coulissements)

Fig. VI-2 - Intra-cratonic depression - "fontanels"
Foreland collapse-holes - "funnels"
(G. Z. 1985.)

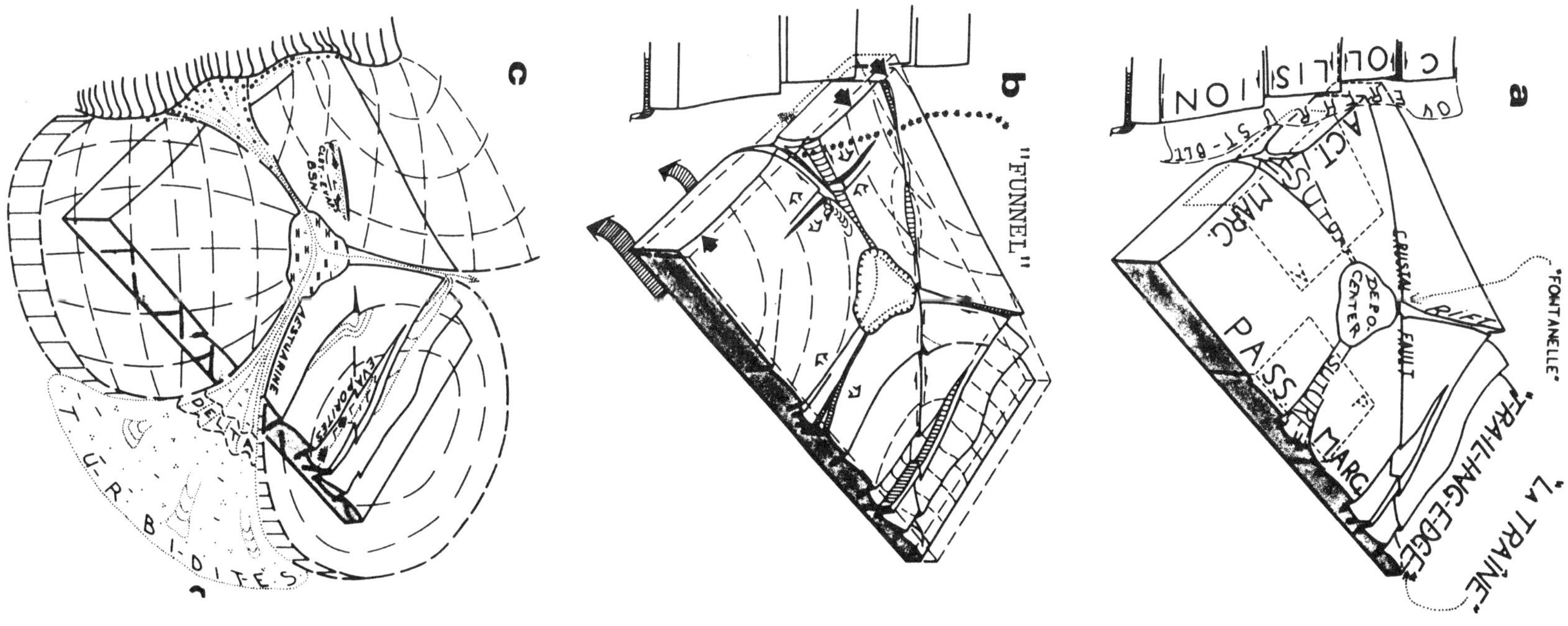

Fig. VI-2 - Dépressions intracratoniques - "fontanelles"
Effondrements d'avant-pays - "entonnoirs"
(G.Z.)

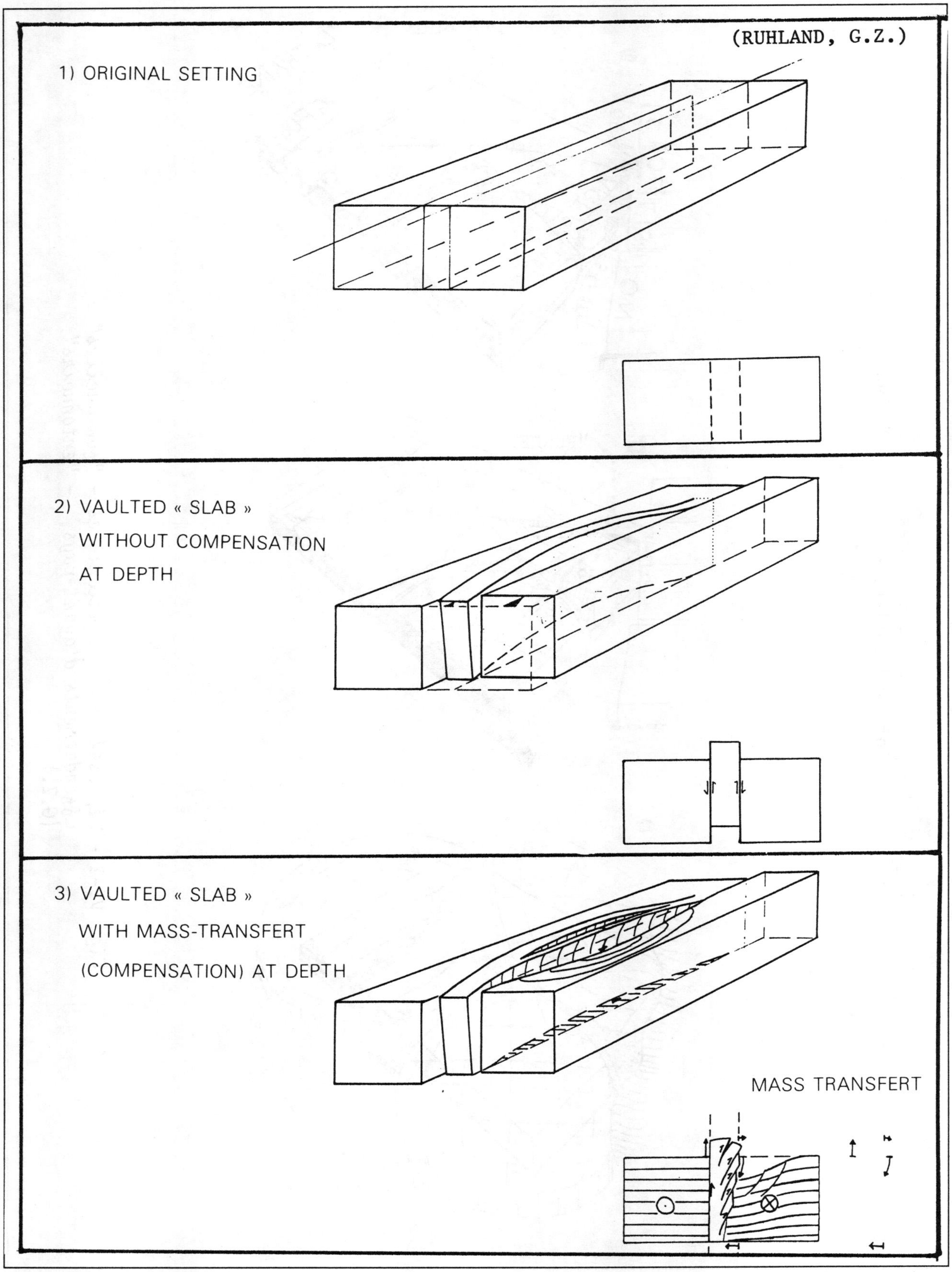

Fig. VI-3 - Shear-Corridors, Uplifts and Basins

- *Couloirs de cisaillement - Voussures - Bassins*

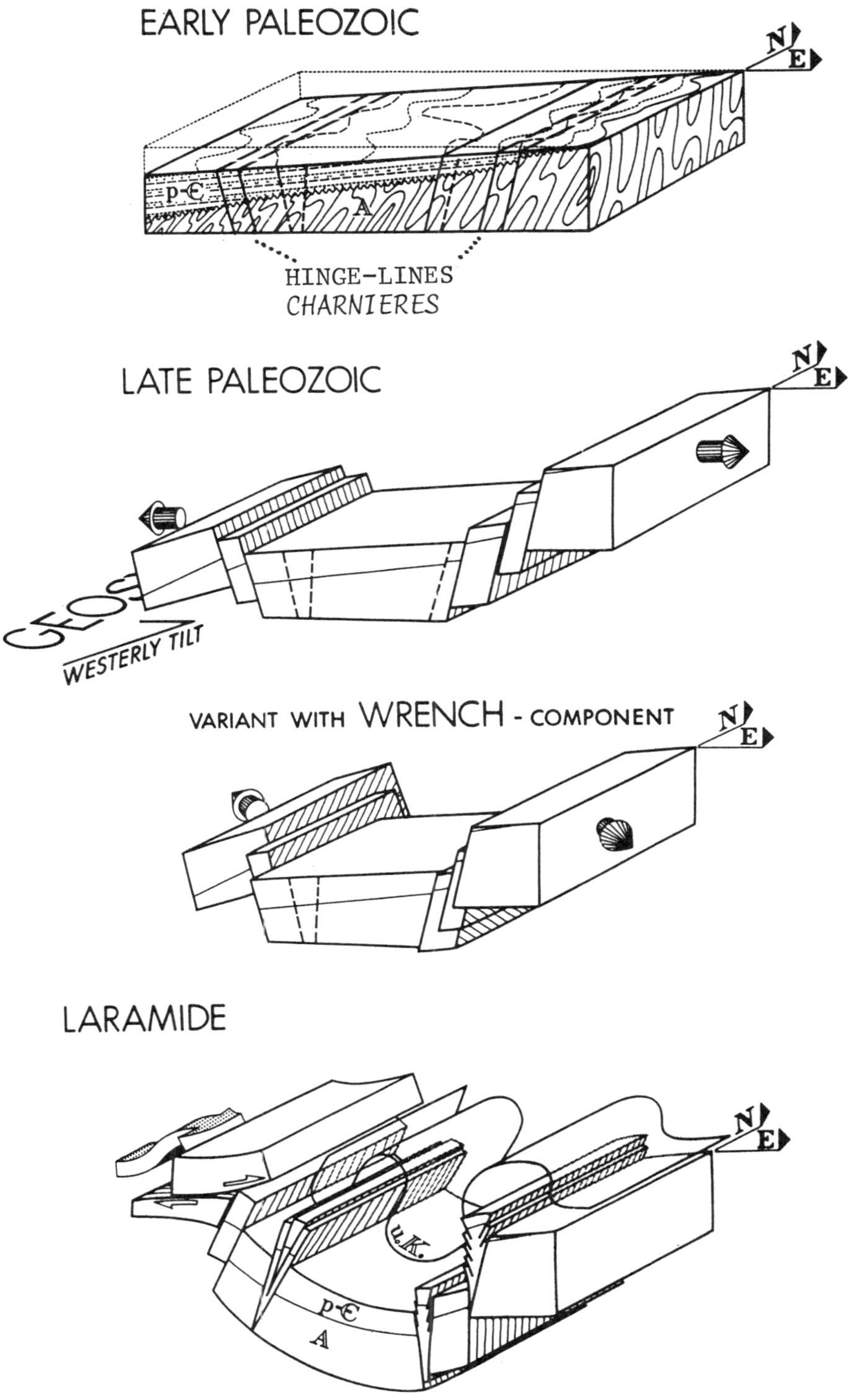

This exemple is taken from the Liard Basin, (n.e.B.C.) N.W. CANADA
Cet exemple est tiré du Bassin de Liard (n.e.B.C.) N.O. CANADA

See also/cf. : Fig. : VI-13 & VII-7

(G.Z. 1982.)

Fig. VI-4 - Hinge-lines & Peripheric Rifts
- Zones de charnière et Rifts périphériques

Lineament-Block Tectonics: Williston-Blood Creek Basin

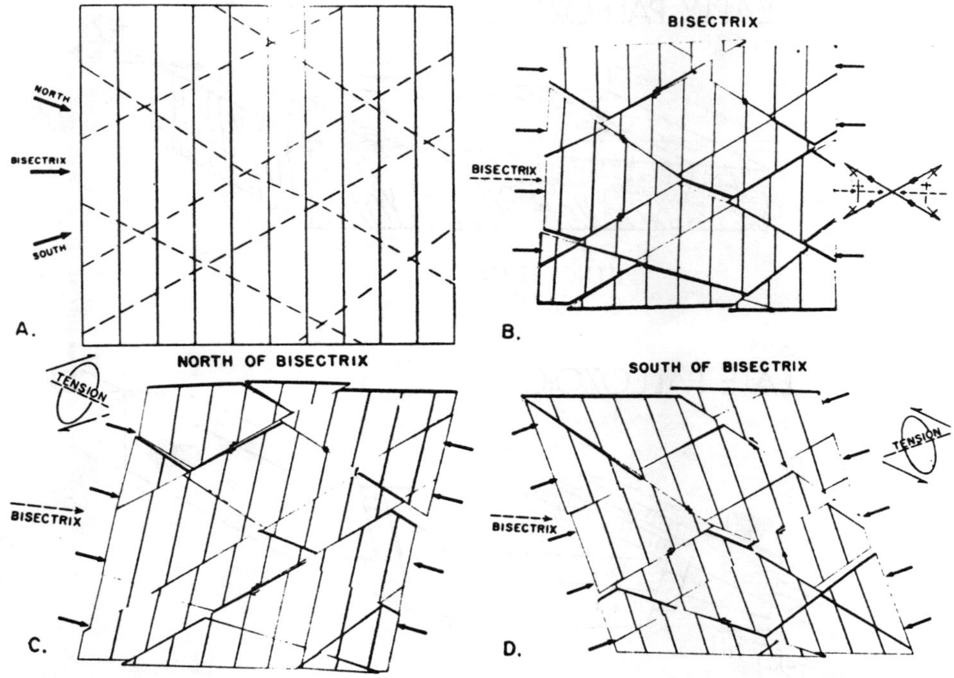

Fig. 7—Cardboard model studies: **A**, initial model; dashed lines are precut weakness zones; solid lines are displacement indicators; arrows represent three compression experiments; **B**, bisectrix direction compression; **C**, north of bisectrix compression; **D**, south of bisectrix compression.

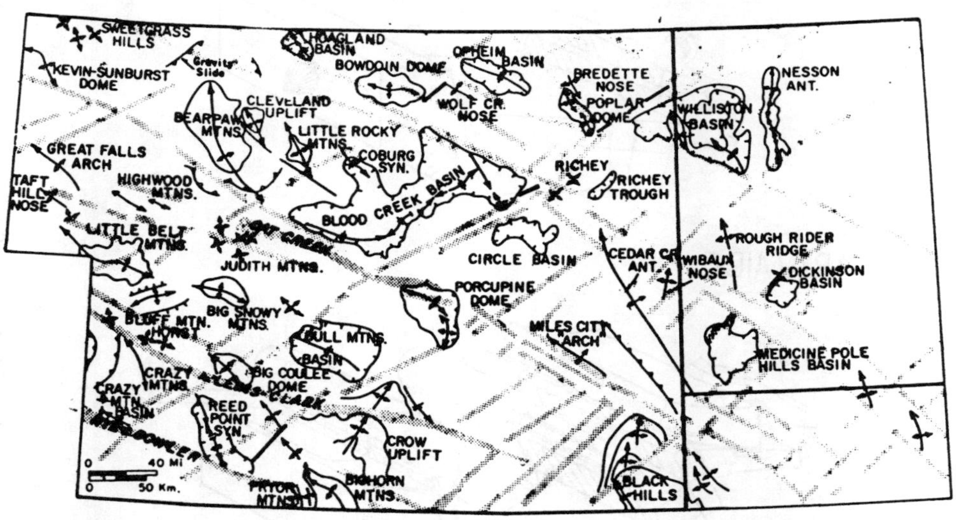

Fig. 6—Comparison of block framework with structural features of Williston–Blood Creek basin area.

Fig. VI-5 - "Chessboards"
(THOMAS)

Fig. VI-5 - Damiers
(THOMAS)

LEGEND FIGURES VI-6, VI-7, VI-8, VI-9

A. Fig. VI-6 : Early-mid PROTEROZOIC (Hudsonian Orogeny)

AFFB	: Amer Fold-Fault Belt	OB	: Ozark Block
BP	: Bear Province	SP	: Superior Province
CB	: Colorado Block	SIP	: Slave Province
ChP	: Churchill Province	WB	: Wisconsin Block
MB	: Montana Block	WP	: Wyoming Province
NGP	: Nain-Greenland Province		

B. Fig. VI.7 : Late PROTEROZOIC (Grenville Orogeny)

AP	: Athabasca Plate	ND	: Nashville Dome
AU	: Adirondack Uplift	NEG	: North-East Greenland "Plate"
BB	: Borden Basin	OBR	: Ottawa-Bonnechère Rift
CA	: Cincinnati Arch	RR	: Reelfoot Rift
LU	: Llanos Uplift	TB	: Thule Basin
MA	: Minto Arch	TP	: Thelon Plate
MCR	: Mid-Continent (Keweenawan) Rift	WR	: Wichita Rift
MMR	: Mid-Michigan "Rift"		
NB	: Nipigon Basin		

C. Fig. VI.8 : PALEOZOIC Orogenies

BA	: Boothia Arch	OB	: Ouachita Orogenic Belt
CCU	: Cedar Creek Uplift	38 F.Z.	: 38th Parallel Fault Zone (Kentucky F.Z.)
MB	: Marathon Orogenic Belt		

D. Fig. VI.9 : MESOZOIC - TERTIARY

BHU	: Black Hills Uplift	NA	: Nesson Anticline
CA	: Cambridge Arch	NEC	: North-East Channel
ChA	: Chardon Arch	SGA	: Sweetgrass Arch
CCA	: Cedar Creek Anticline	SLCh	: Saint Lawrence Channel
FC	: Flemish Cap	SLR	: Saint Lawrence Rift
FMF	: Franklin Mountain Foldbelt	WP	: Wyoming Province
GF	: Great Falls tectonic zone		
LCL	: Lewis-and-Clark Lineament		

The basic scheme of these maps follows (WISE, 1963 - MUEHLBERGER, 1980 and ZOLNAI, 1986)

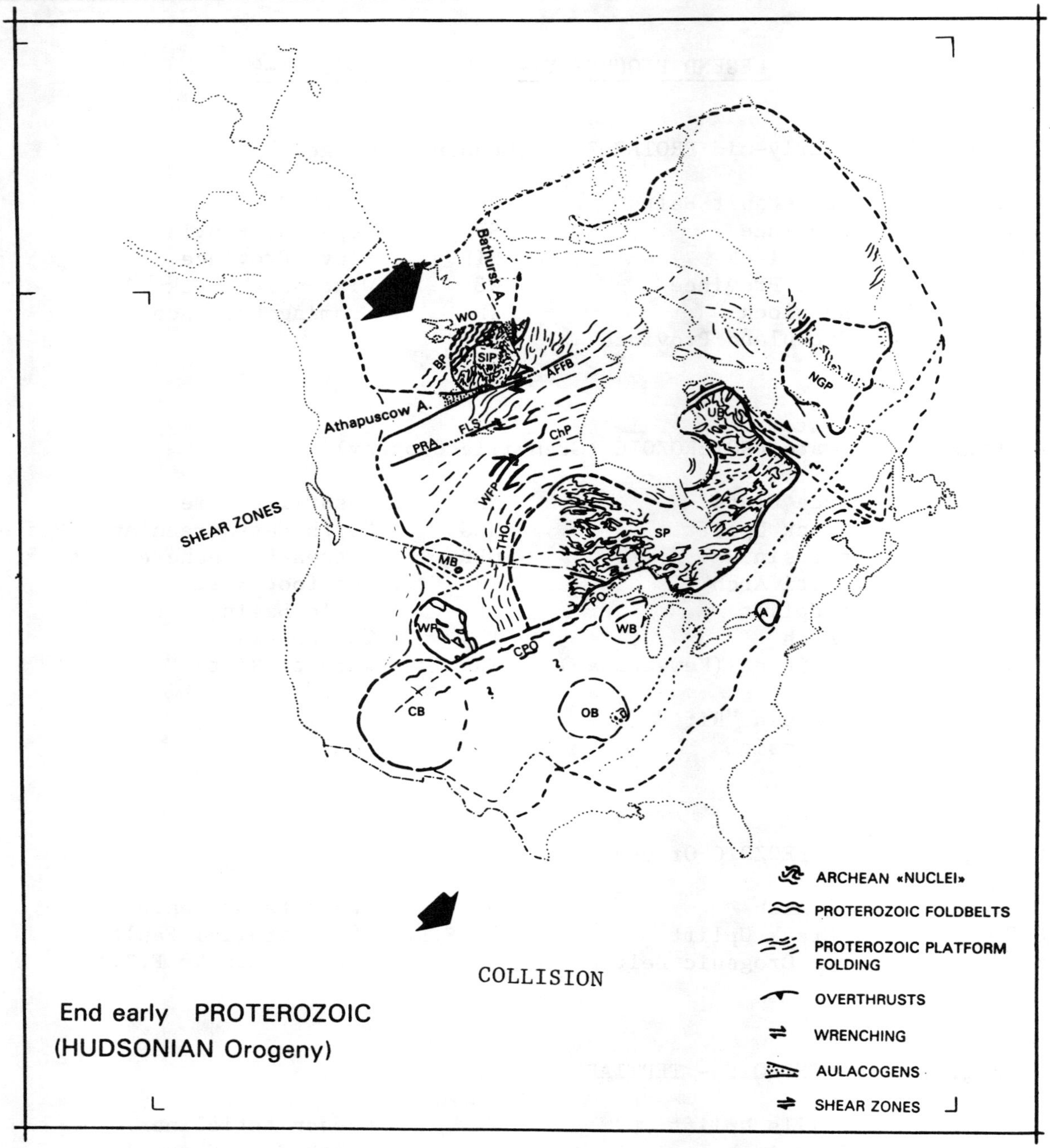

Fig. VI-6 - Evolution of the North American Continent : Mid-Proterozoic (G.Z.)

- *L'évolution du Continent Nord-Américain : Protérozoïque moyen* (G.Z.)

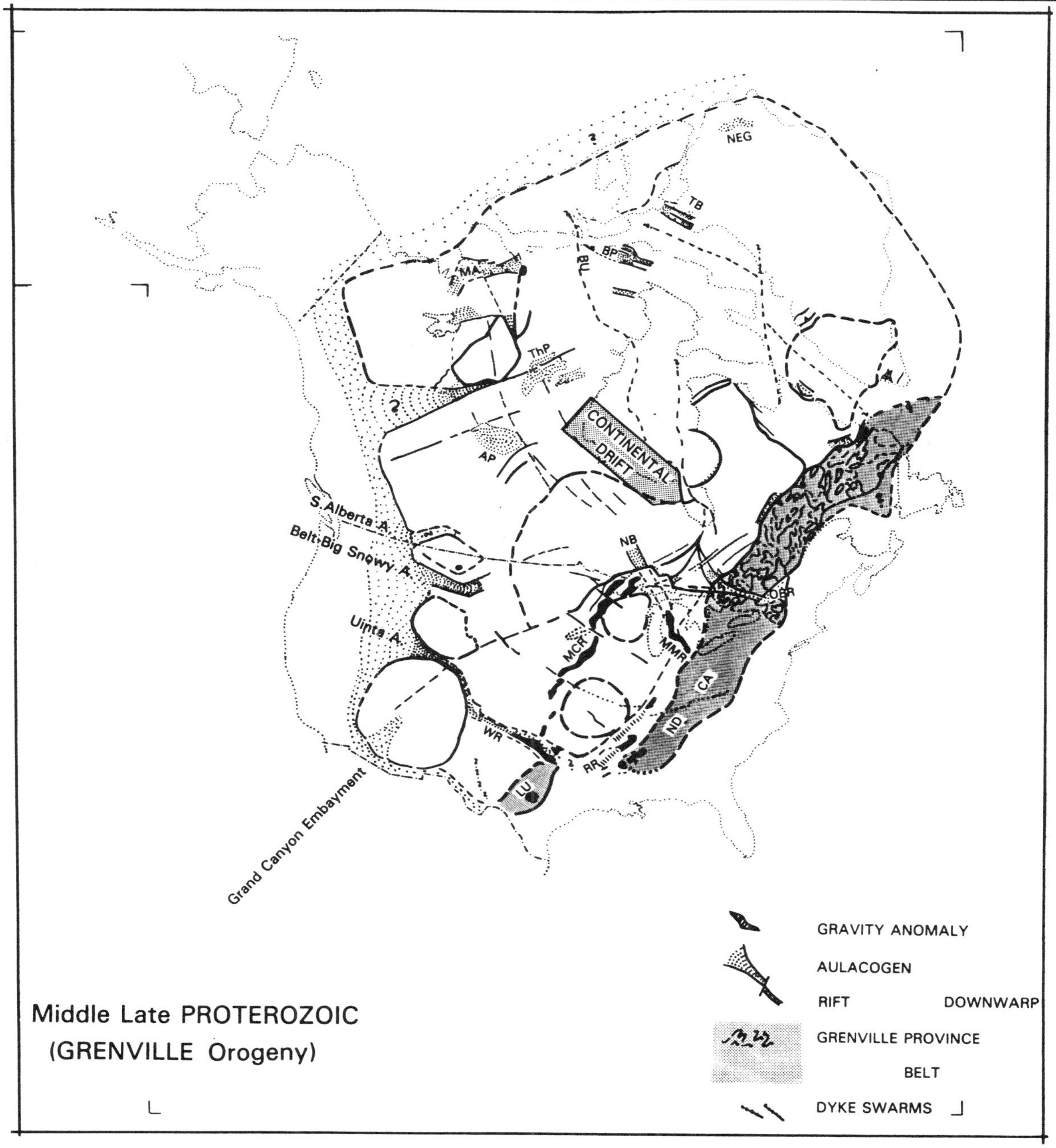

Fig. VI-7 - Evolution of the North American Continent : Late-Proterozoic (G.Z.)

- *L'évolution du Continent Nord-Américain : Protérozoïque supérieur (G.Z.)*

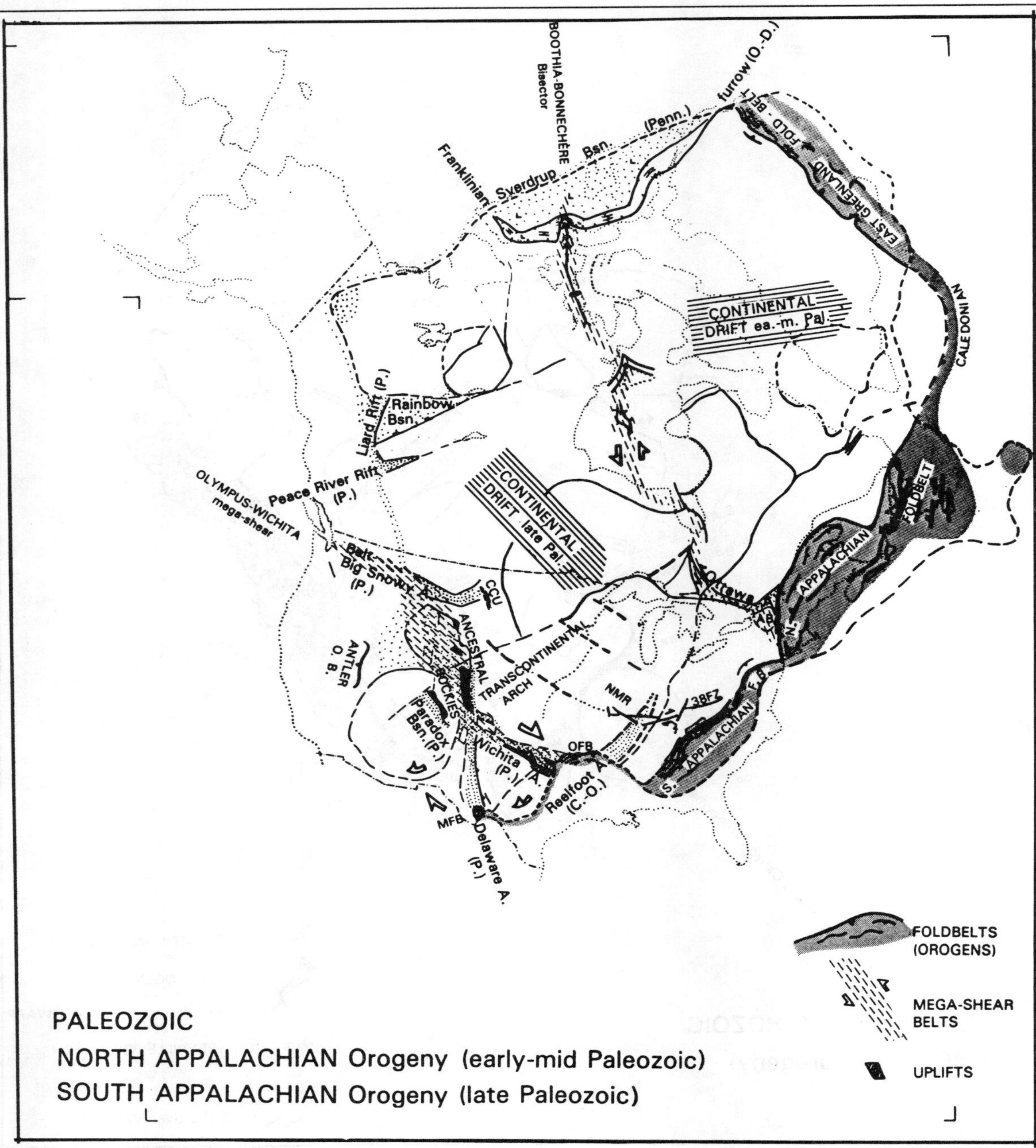

Fig. VI-8 - Evolution of the North American Continent : Paleozoic (G.Z.)

- *L'évolution du Continent Nord-Américain : Paléozoïque* (G.Z.)

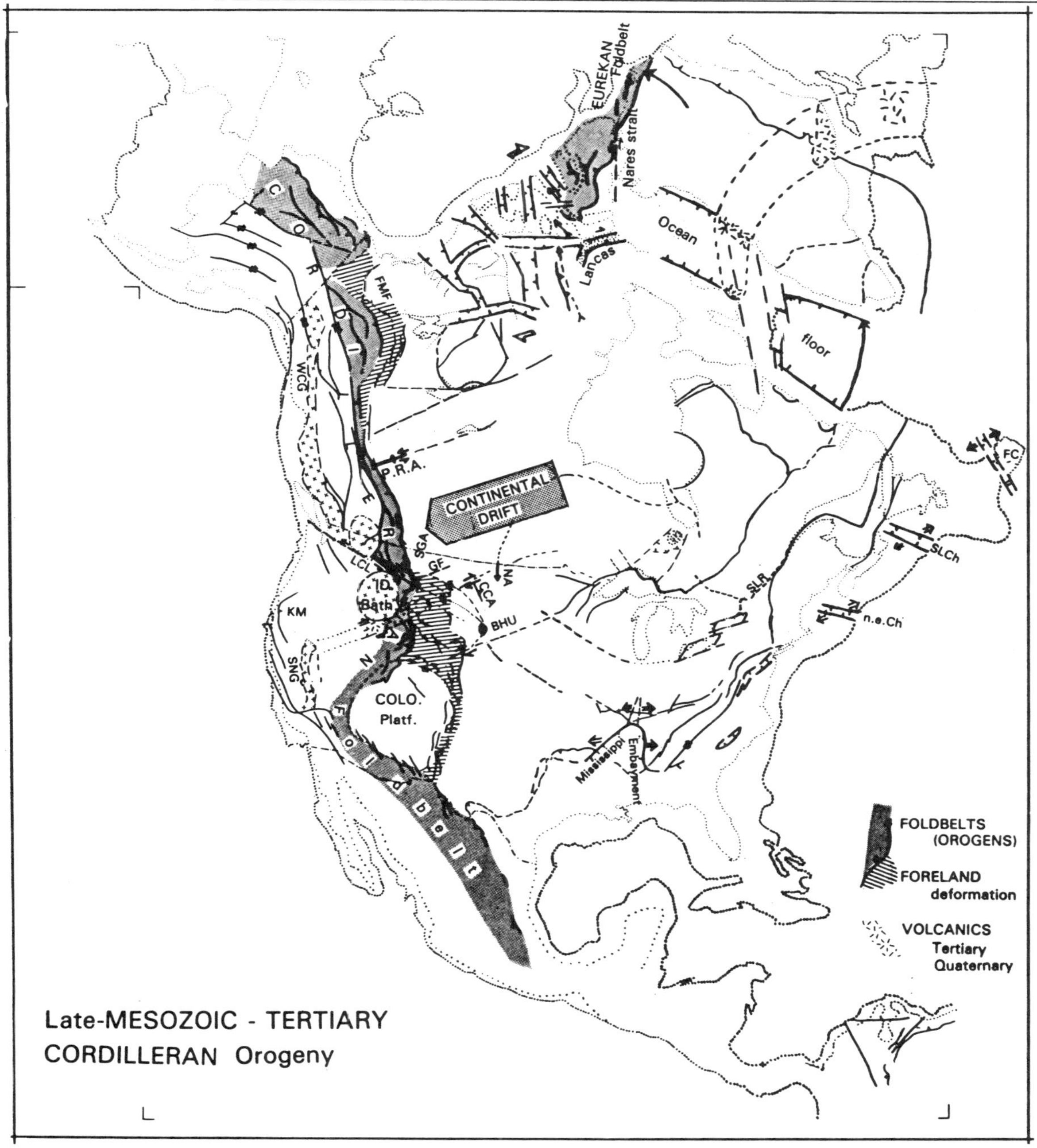

Fig. VI-9 - Evolution of the North American Continent : Mesozoic-Tertiary (G.Z.)

- *L'évolution du Continent Nord-Américain : Mésozoïque-Tertiaire (G.Z.)*

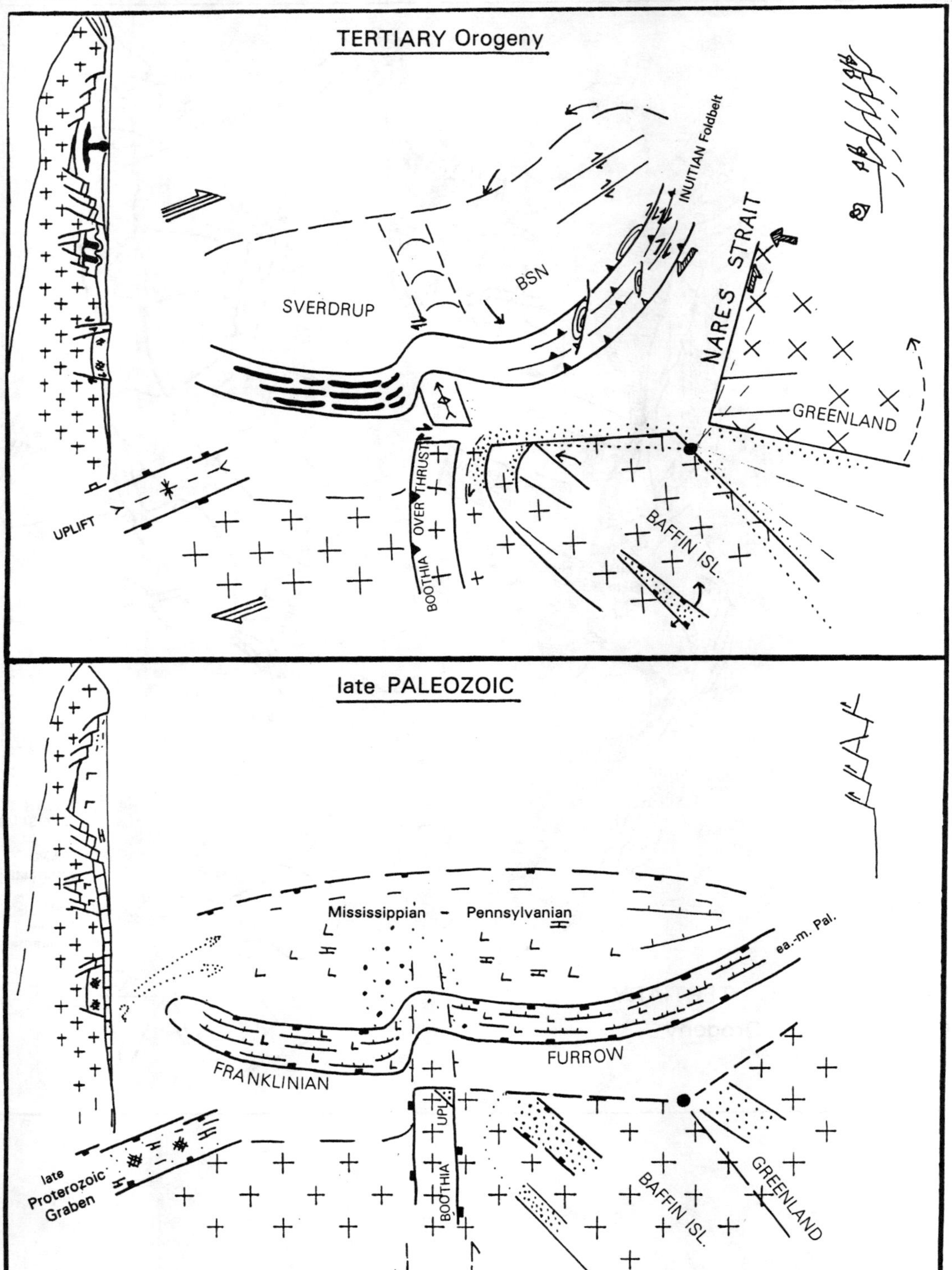

Fig. VI-10 - A passive-margin orogeny : Artic Islands/Canada

- *Une zone orogénique "passive" : Iles Arctiques, Canada*

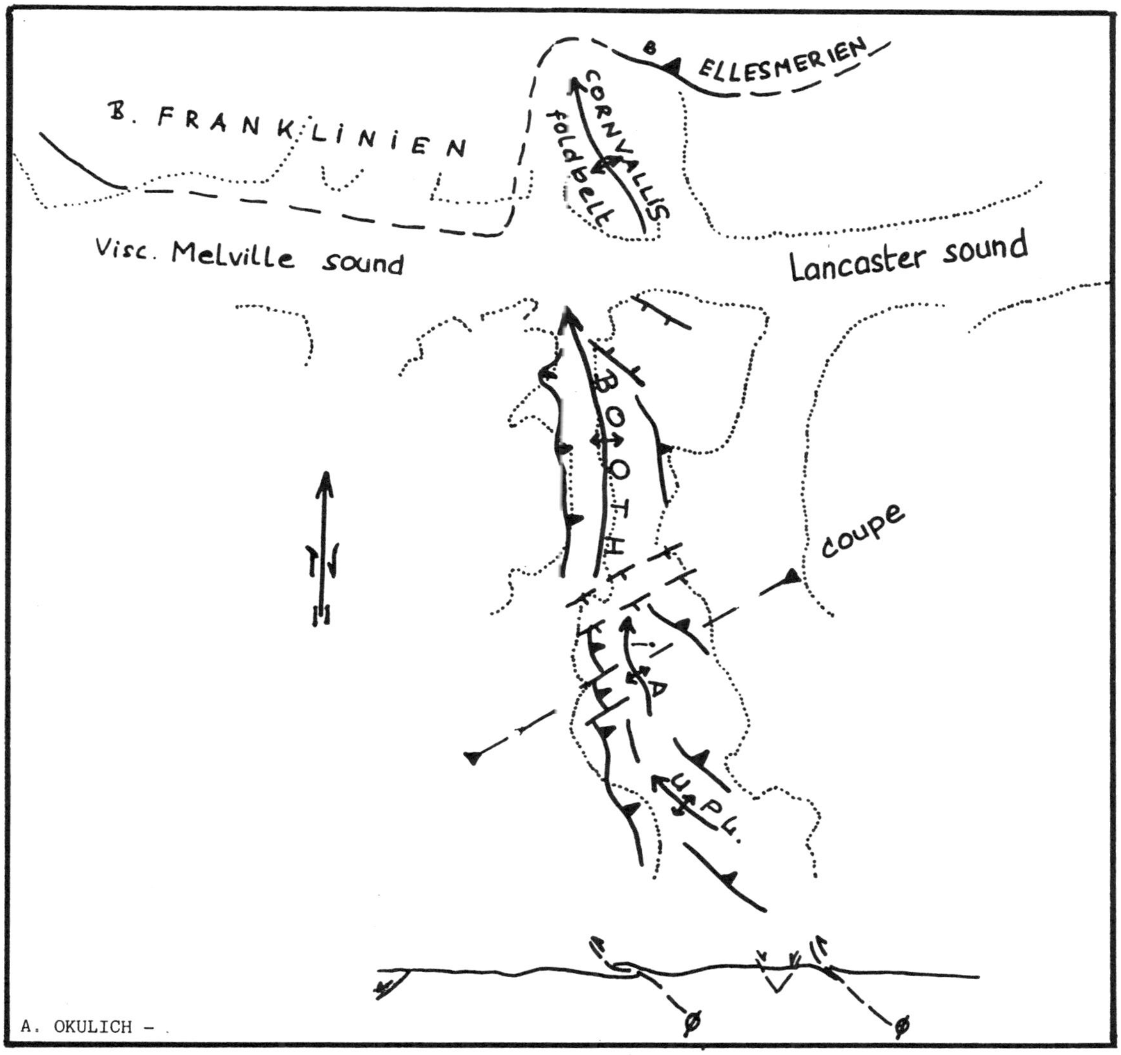

(From/d'après: OKULICH et al. - 1986.)

BOOTHIA uplift, Arctic Canada.

Fig. VI-11 - A passive-margin orogeny : Artic Islands/Canada

Fig. VI-11 - Une zone orogénique "passive" : Iles Arctiques, Canada

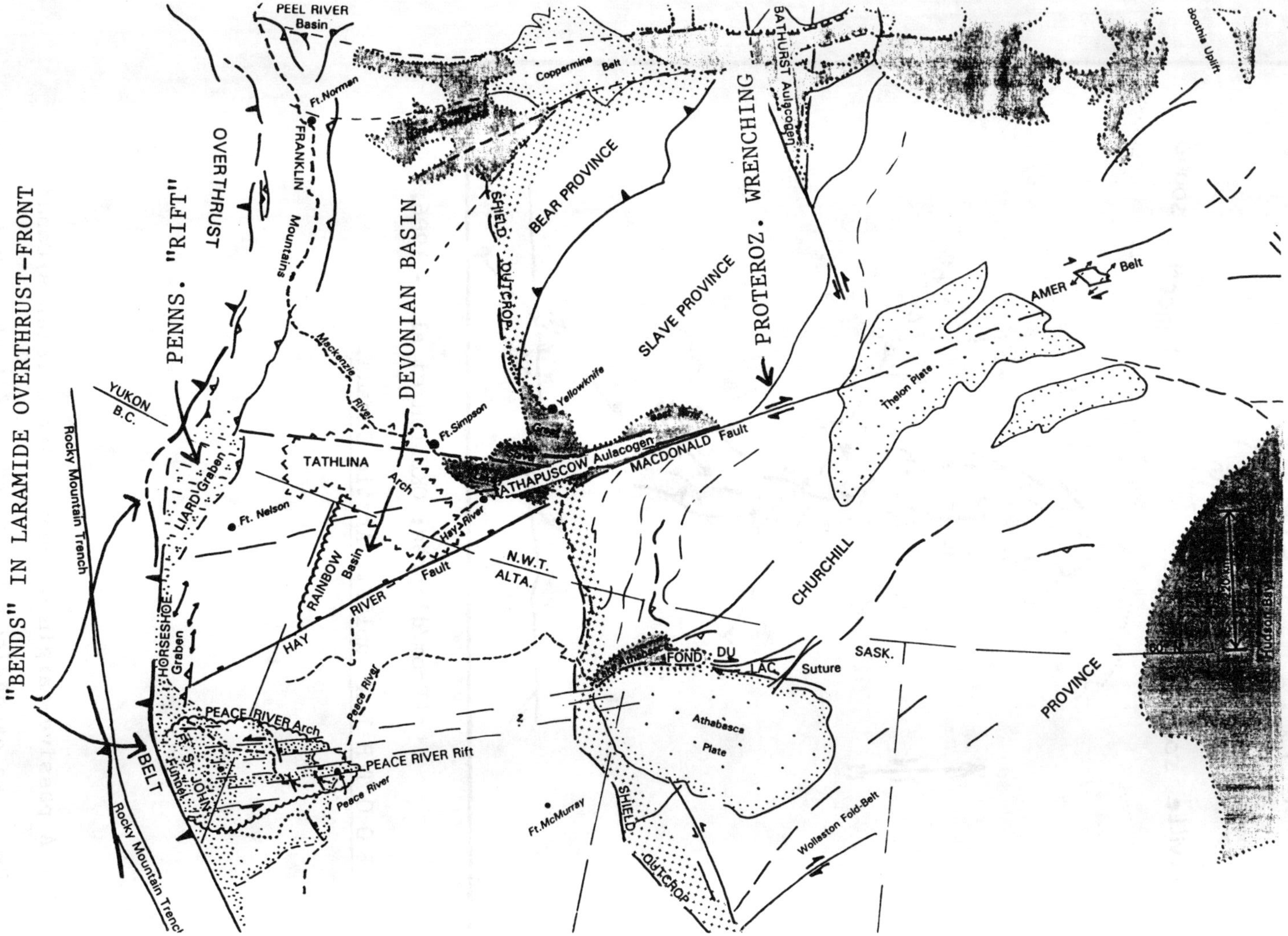

Fig. VI-12 - Liard - Rainbow - Athapuscow - Amer belt - Peace River - Athabasca Lake (N.W. Canada)

- *Liard - Rainbow - Athapuscow - Amer "belt" - Rivière-la-Paix - Lac d'Athabasca (N.W. Canada)*

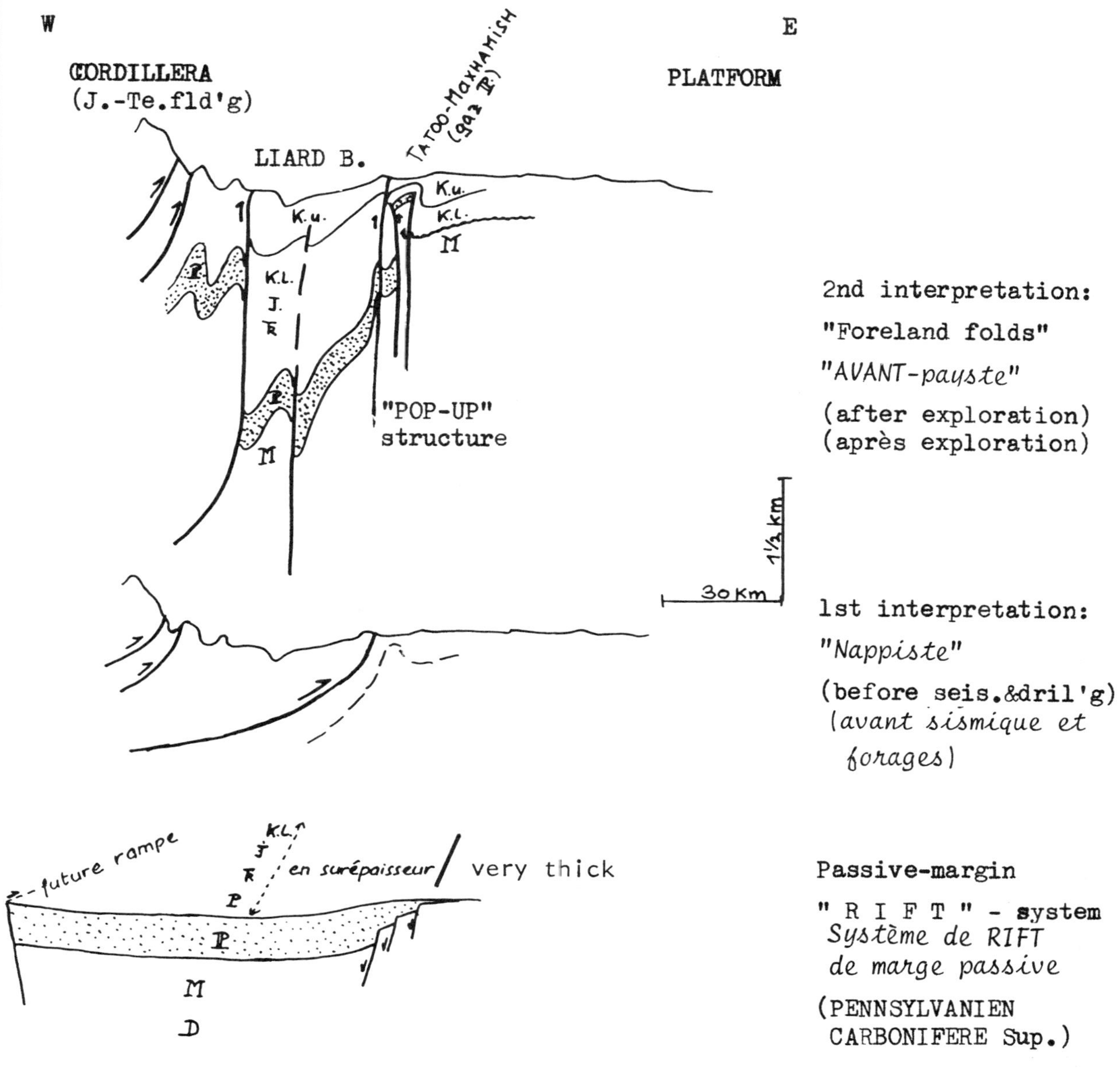

Fig. VI-13 - Liard Basin
(G.Z.)

Fig. VI-13 - Bassin de Liard
(G.Z.)

A 75 km right-lateral wrenching has occured during the life-span (lower-mid Proterozoic) of the ATHAPUSCOW Aulacogen. For position see Fig. VI-12.

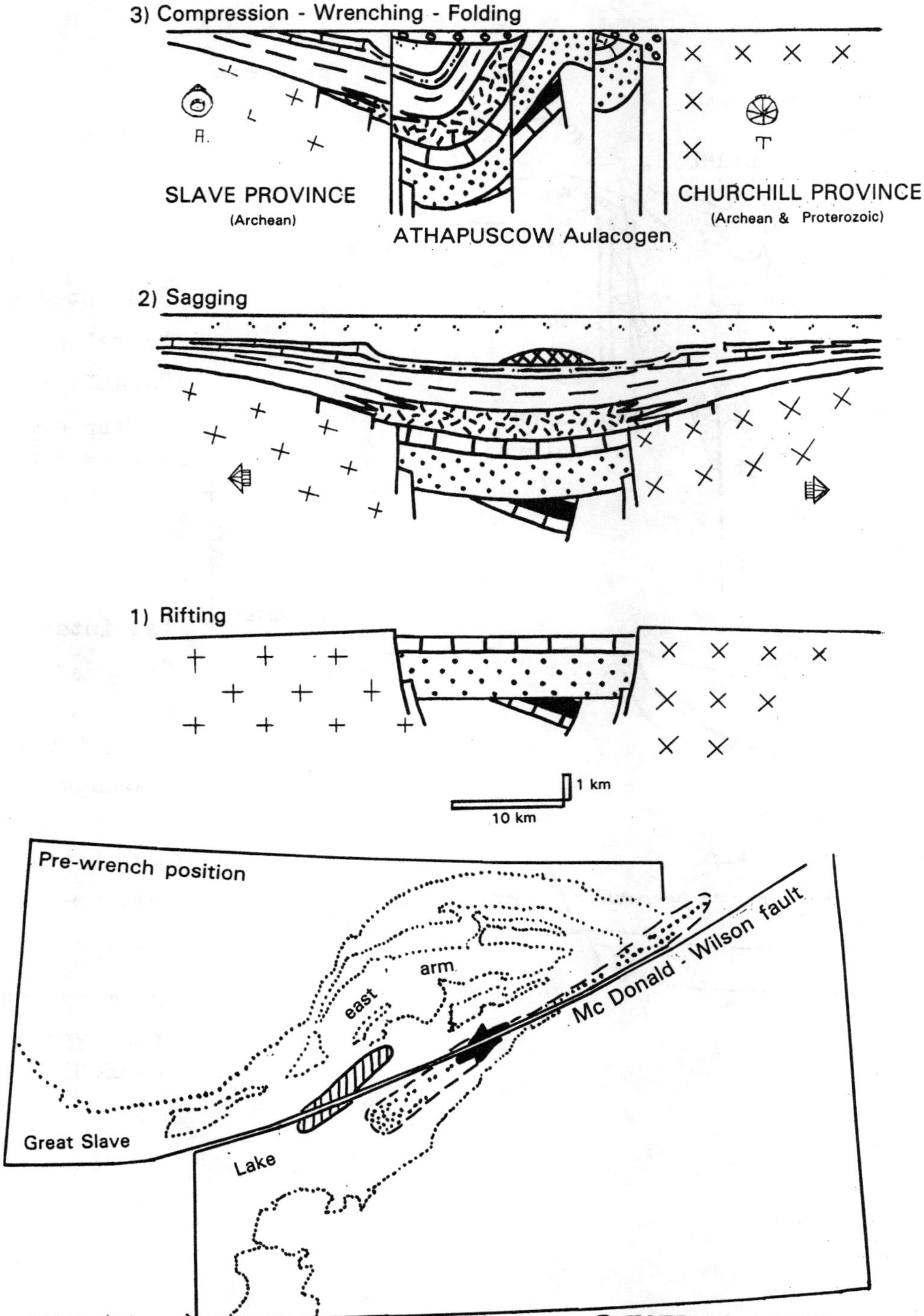

After/d'après: P.HOFFMAN - 1977.

Fig. VI-14 - Athapuscow aulacogen (HOFFMAN)

- *Aulacogène d'Athapuscow* (HOFFMAN)

Un coulissage dextre de 75 km a eu lieu pendant la formation et la déformation de l'Aulacogène d'ATHAPUSOW. Cf. Fig. VI.12.

From/de: J.PATTERSON - 1986

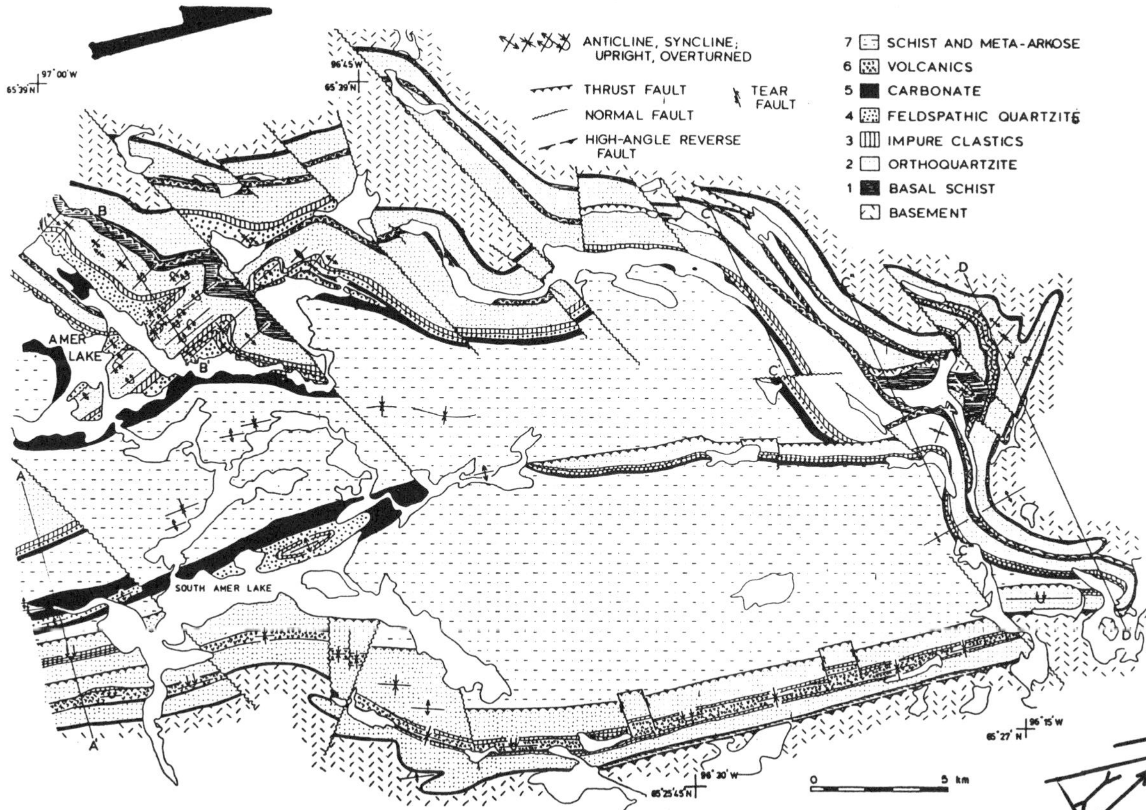

Geological map of the eastern end of the Amer Belt.

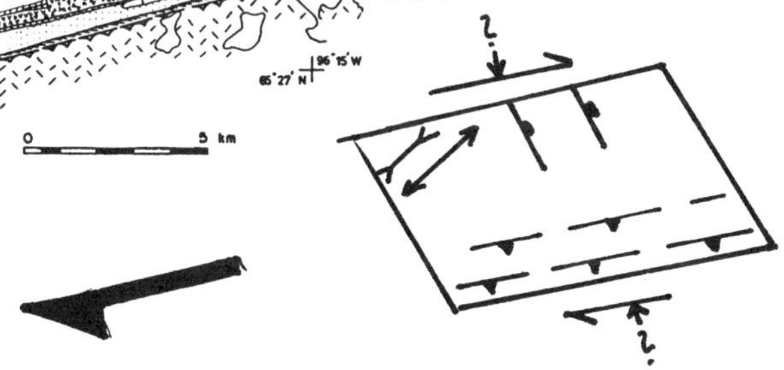

It is tempting to interpret this rhomb-shaped basin as being due to a right-lateral wrenching; especially since it falls in line with the 'East Arm system':Fig.VI-12 & 14

On peut interpréter ce bassin de forme losangique comme étant dû à un coulissage dextre, d'autant plus qu'il se trouve dans la prolongation du système de décrochement du Lac des Esclaves ; cf. VI-12, 14.

Fig. VI-15 - Amer Belt (PATTERSON)

Fig. VI-15 - Amer Belt (PATTERSON)

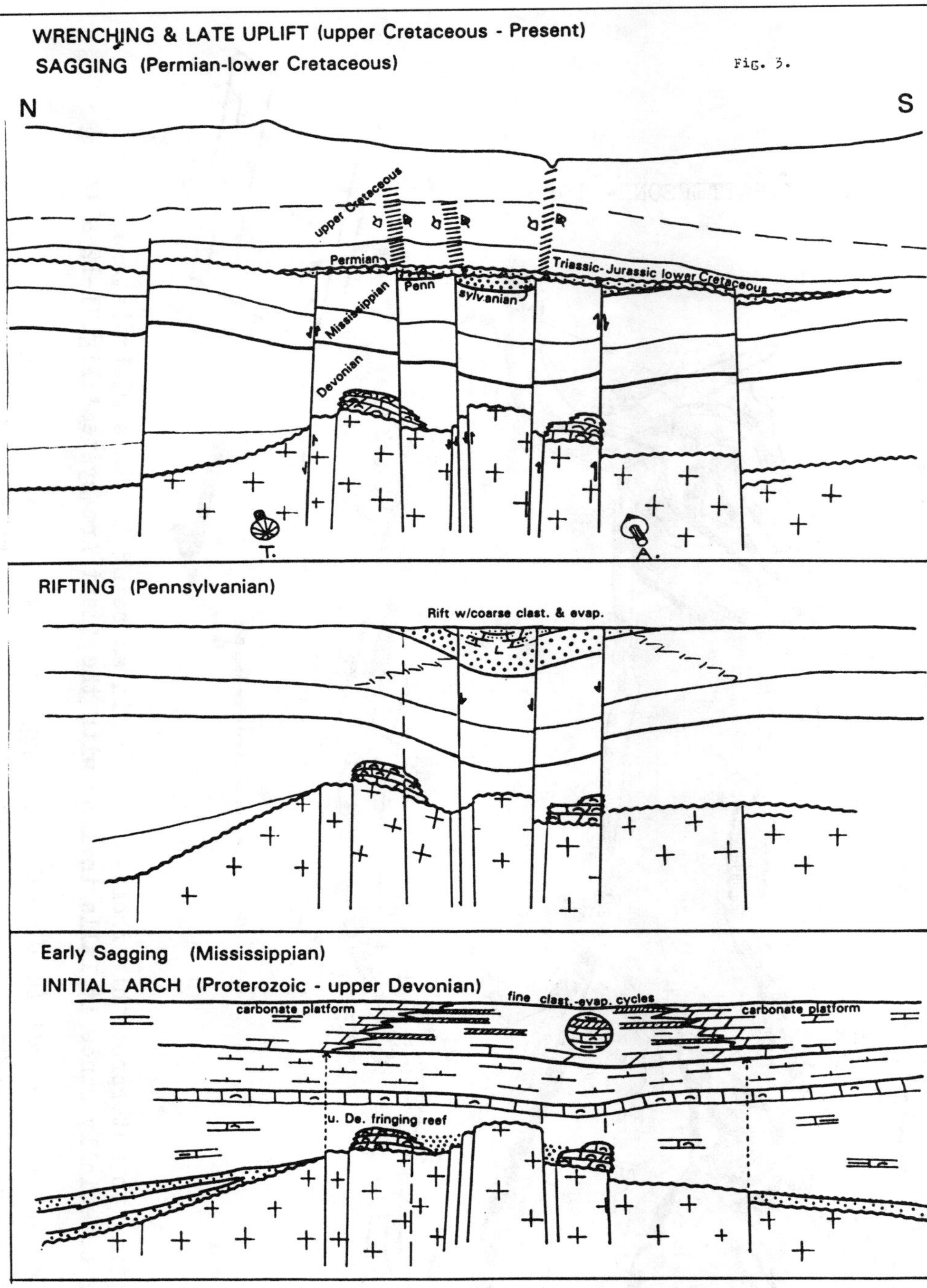

Fig. VI-16 - Peace River Arch/Trough
(G.Z.)

Fig. VI-16 - L'Arche/Effondrement de "Peace River"
(G.Z.)

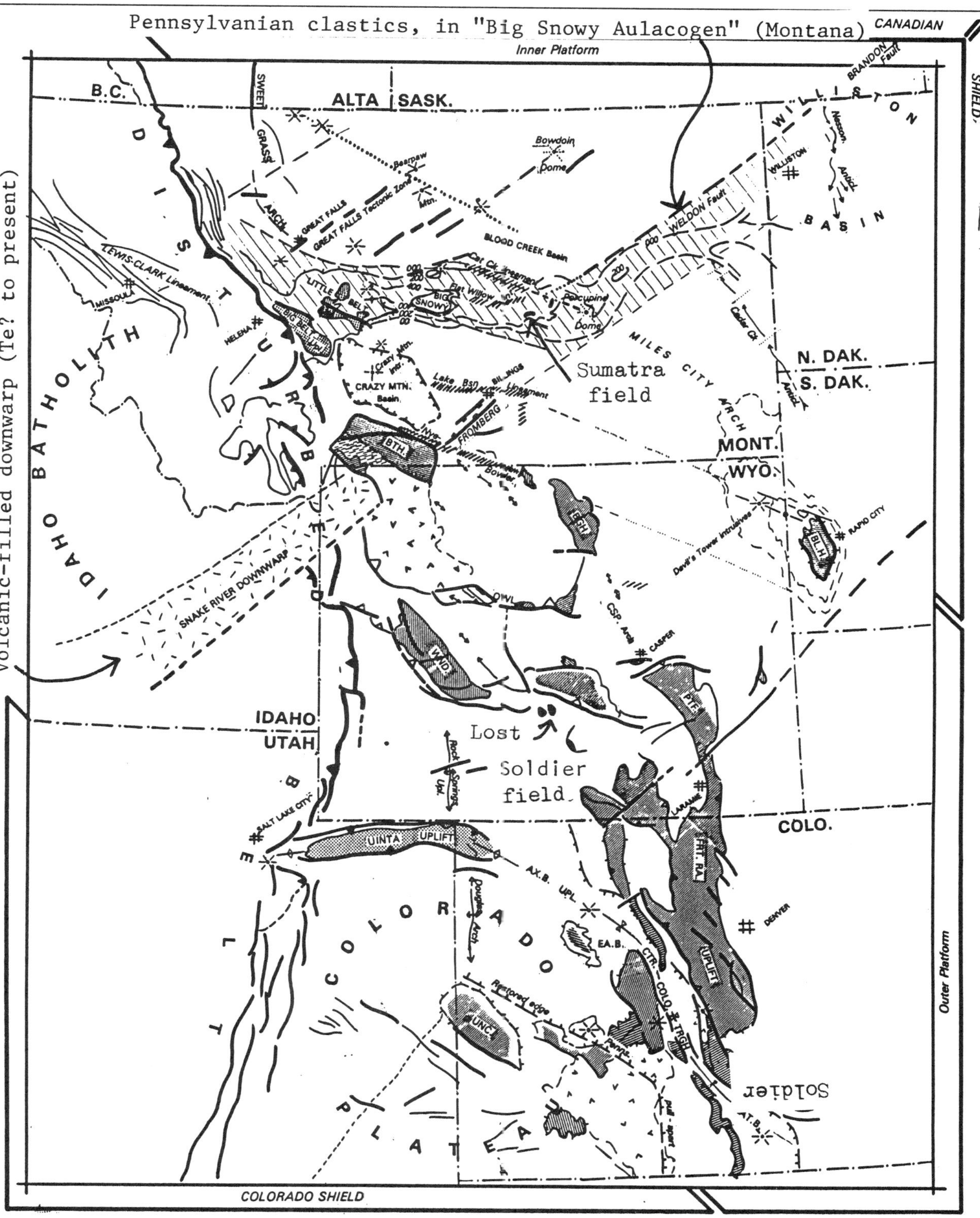

Fig. VI-17 - Belt - Big Snowy - Black Hills uplifts - Snake River - Billings Weldon faults (W. USA)

- *Belt - Big Snowy - Black Hills - Snake River - Billings - Weldon (W. USA)*

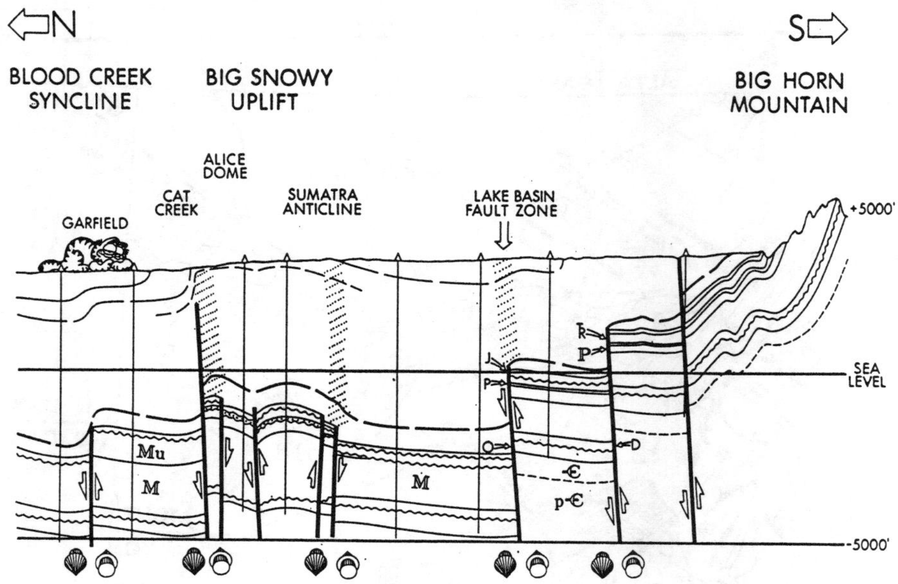

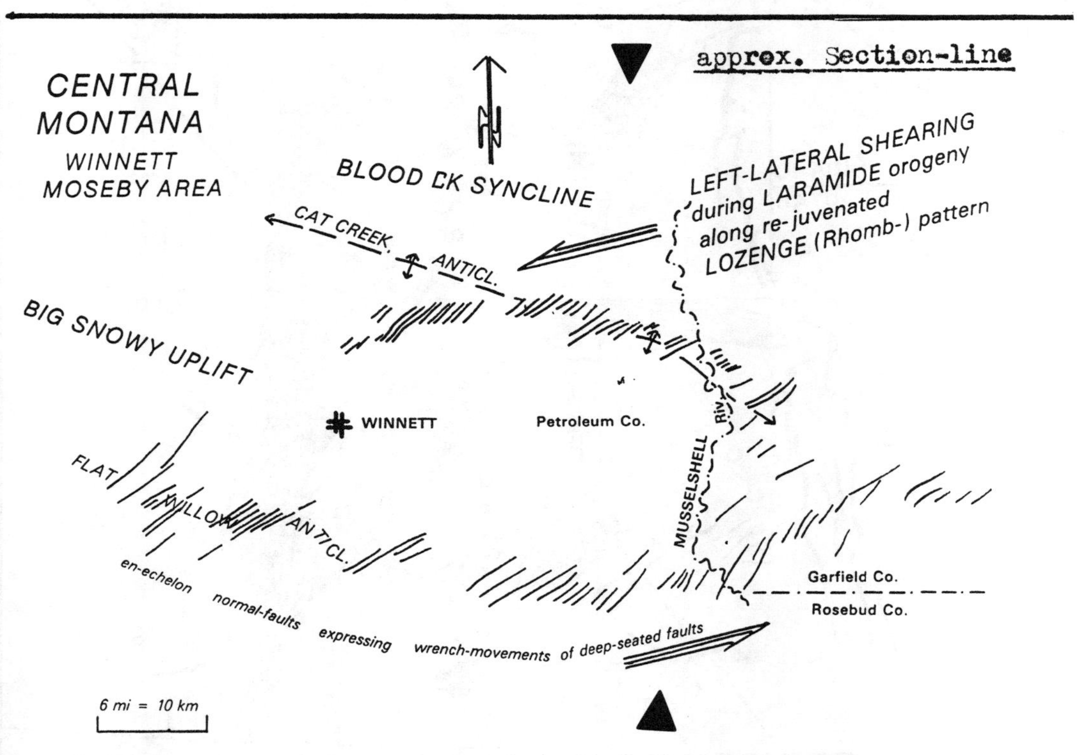

(G.Z.)

Fig. VI-18 - Sumatra - Cat Creek section

- *Coupe de Sumatra-Cat Creek*

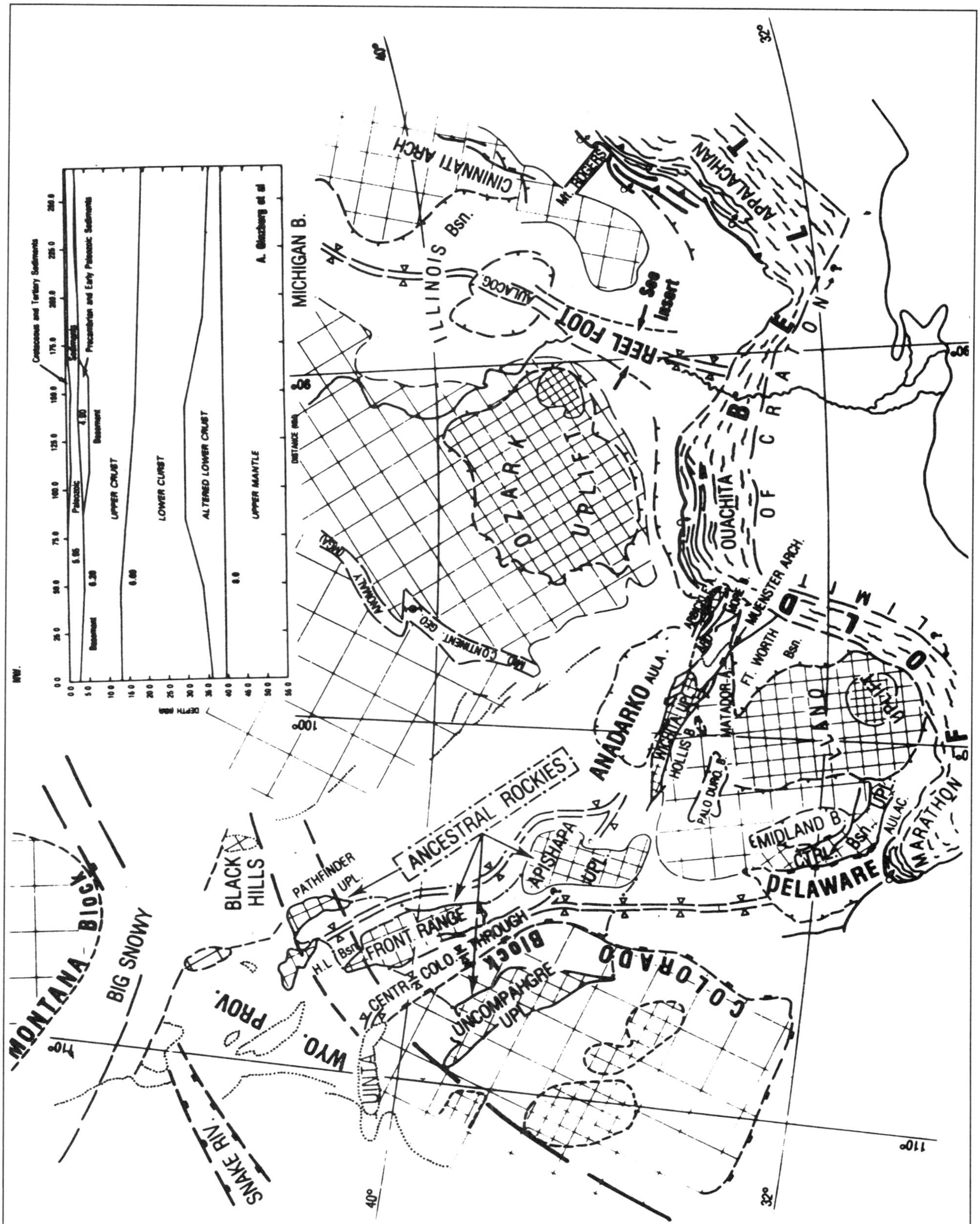

Fig. VI-19 - The "Ancestral Rockies" system, late Paleozoic (W. USA)
(**based** on documents by U.S.G.S.)

- *Le système des "Rocheuses Ancestrales" (Paléozoïque supérieur) (W. USA)*

PROFESSIONAL PAPER 853
PLATE 15-C

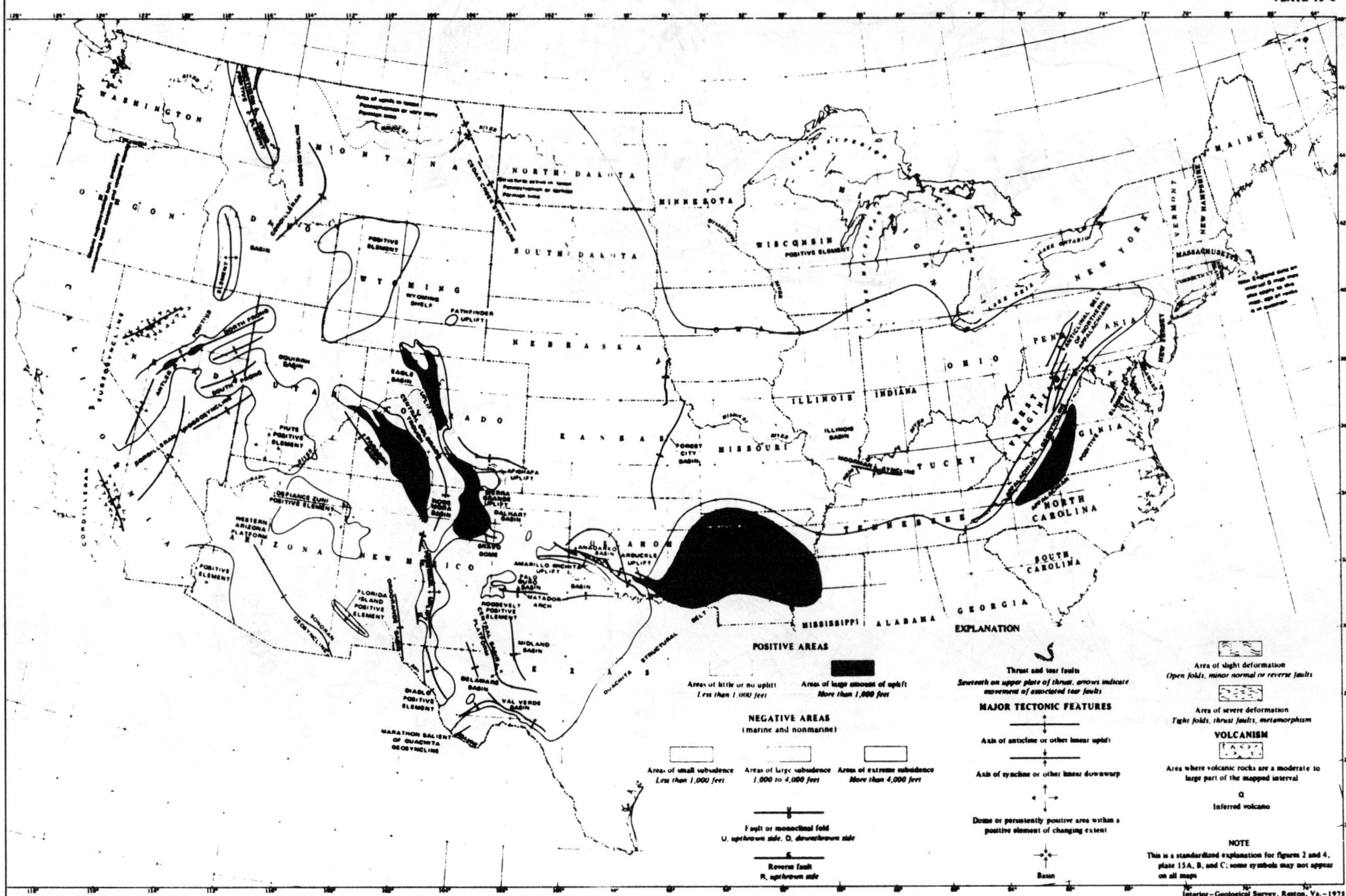

STRUCTURAL FEATURES OF INTERVAL E DETERMINED FROM RESTORED ISOPACH MAP (FIG. 1)

Fig. VI-19/B Structural elements during the Pennsylvanian, U.S.A. - U.S.G.S. 1975.
Eléments structuraux pendant le Carbonifère supérieur.

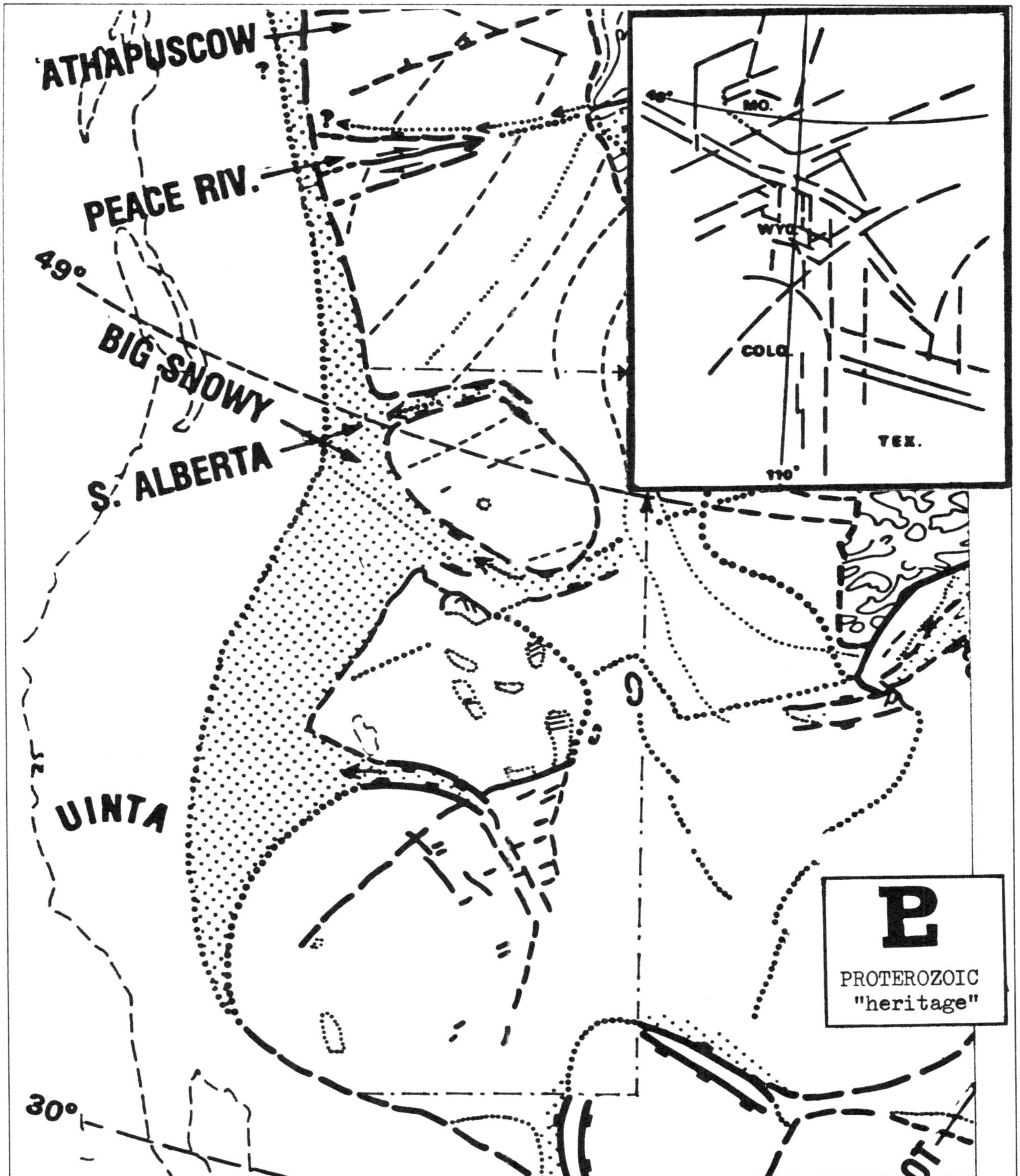

Fig. VI-19/C-D-E-F "ANCESTRAL" vs. "LARAMIDE" Rockies
(G.Z. 1985)

*Des Rocheuses "ANCESTRALES" aux Rocheuses "LARAMIENNES"
(G.Z. 1985)*

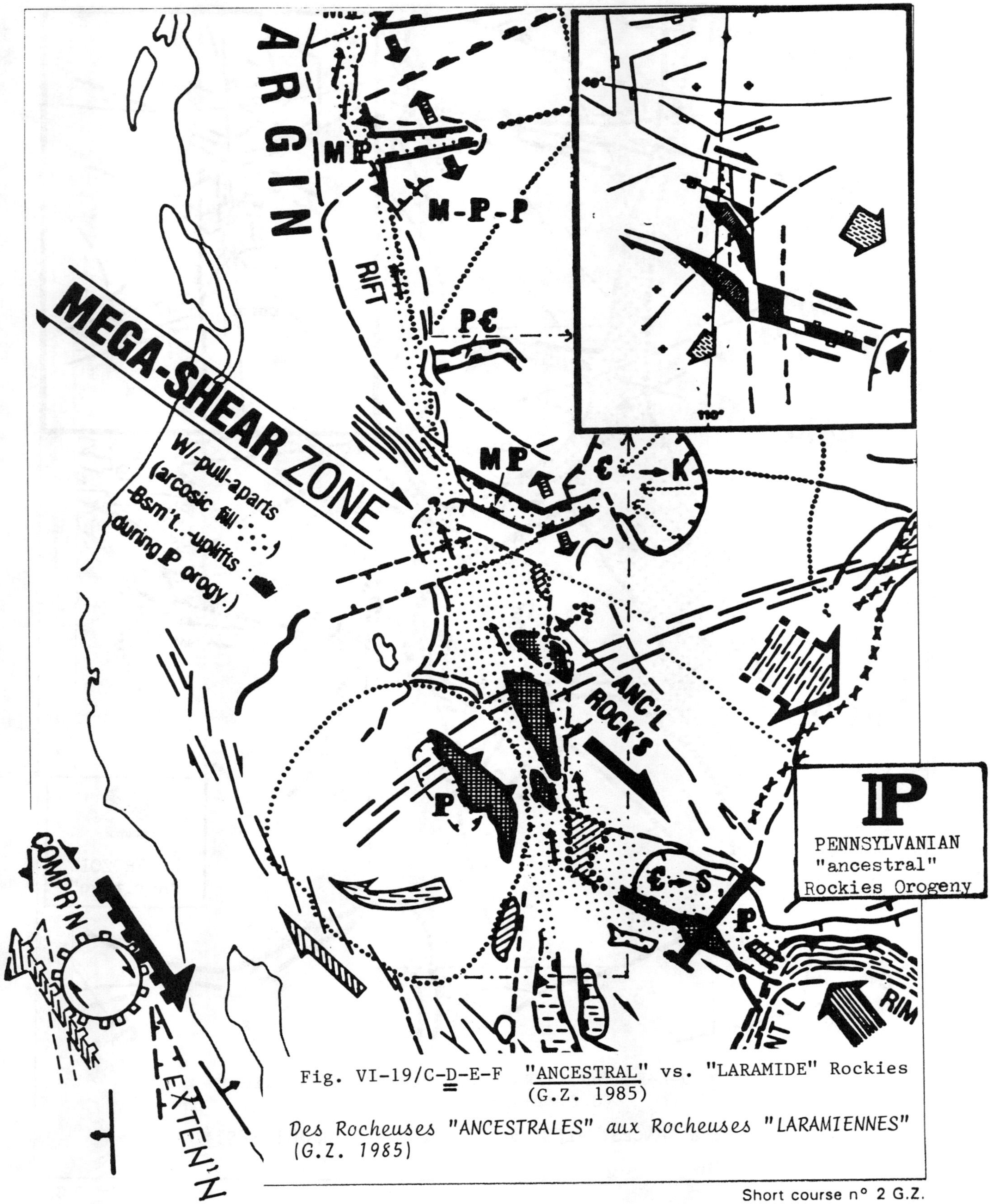

Fig. VI-19/C-D-E-F "ANCESTRAL" vs. "LARAMIDE" Rockies (G.Z. 1985)

Des Rocheuses "ANCESTRALES" aux Rocheuses "LARAMIENNES" (G.Z. 1985)

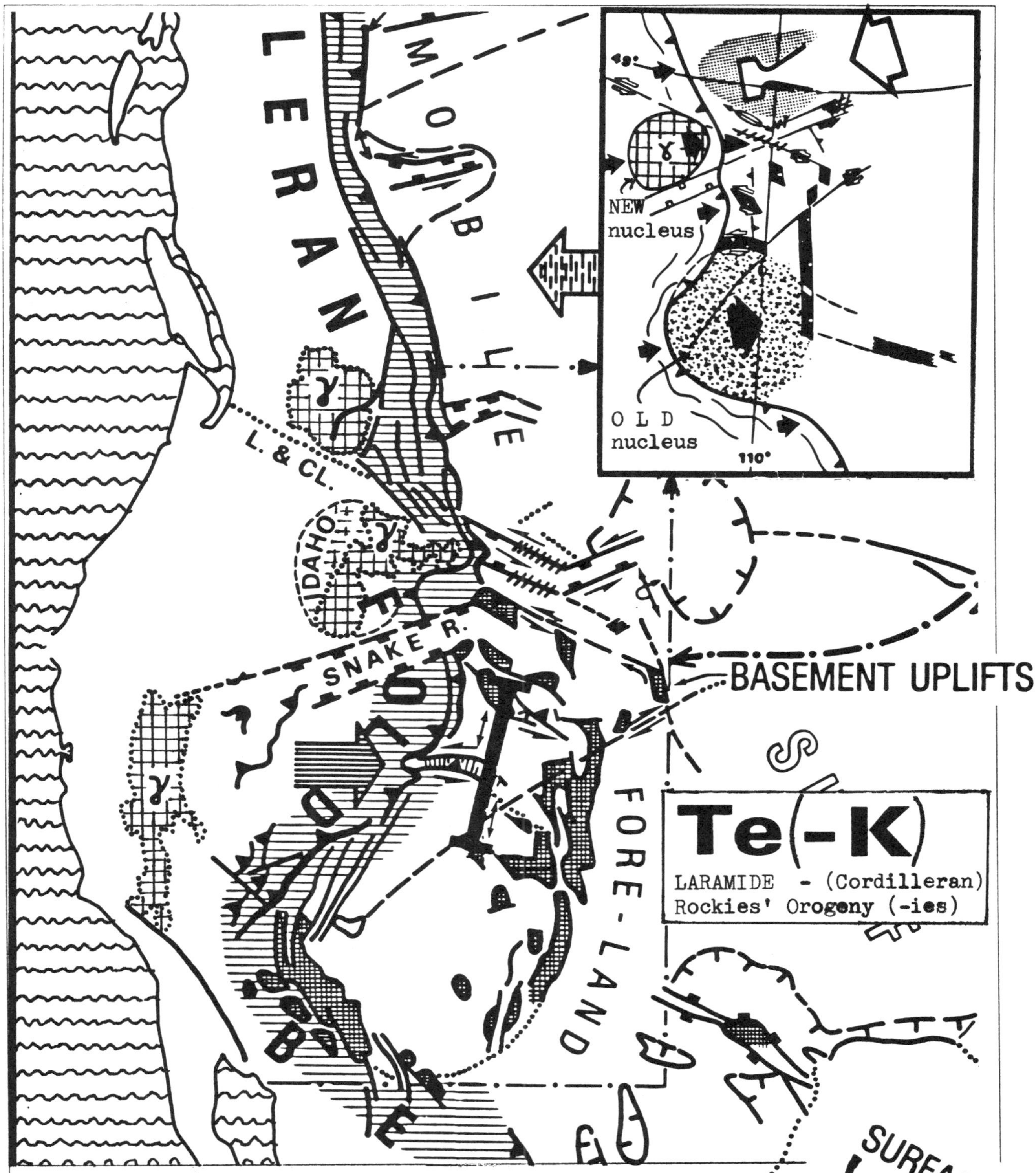

Fig. VI-19/C-D-E-F "ANCESTRAL" vs. "LARAMIDE" Rockies
(G.Z. 1985)

Des Rocheuses "ANCESTRALES" aux Rocheuses "LARAMIENNES"
(G.Z. 1985)

Fig. VI-19/C-D-E-F Des Rocheuses "ANCESTRALES" aux Rocheuses "LARAMIENNES" (G.Z. 1985)

During the Laramide revolution (Paleocene) the basic rhomb-pattern, generated during the late Paleozoic orogenic events by a right-lateral, NW-SE oriented transcontinental mega-shear, has been rejuvenated by a conjugate, compressive wrench-system (NE-SW : right-lateral, NW-SE left-lateral : note the reversal of the sense of wrenching along this direction). The late-Paleozoic (Pennsylvanian : upper Carboniferous) "passive" tectonism has, in this area, remobilized parts of an earlier (mid- to late-Proterozoic ?) structural framework.

The late-Paleozoic wrenching is due, in this interpretation, to a "simple shear" couple ; while the Tertiary wrench-movements correspond to an E-W oriented "pure shear" couple.

L'Orogenèse laramienne (Paléocène) a remobilisé le jeu des blocs rhombiques, façonné par les mouvements tectoniques ("passifs") du Paléozoïque tardif (Pennsylvanien ou Carbonifère supérieur) qui, à leur tour, n'ont fait que reprendre une partie d'un réseau de failles plus ancien encore (Protérozoïque moyen et supérieur ?).

Le couple dextre pennsylvanien, orienté NW-SE, est dû à un cisaillement "simple". -Le réseau compressif conjugué NW-SE (sénestre) et NE-SW (dextre) laramien reflète un cisaillement "pur".- Notons l'inversion du sens des coulissages selon la direction NW-SE, entre les mouvements "ancestraux" et "laramiens".

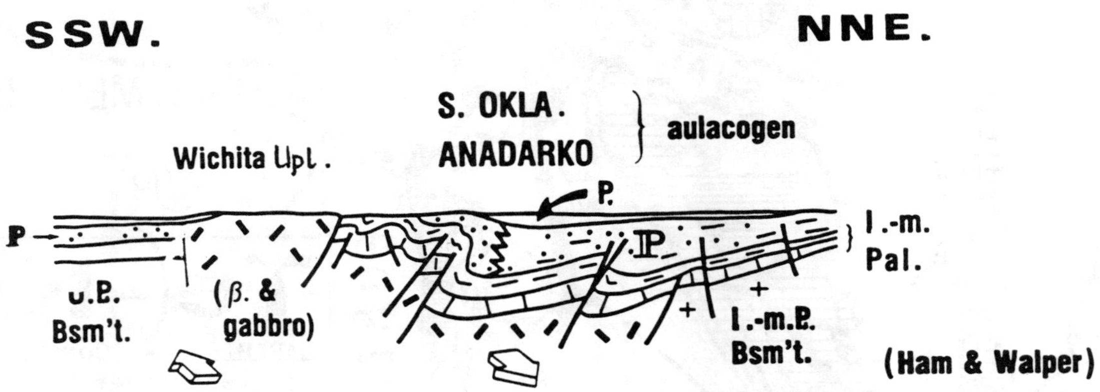

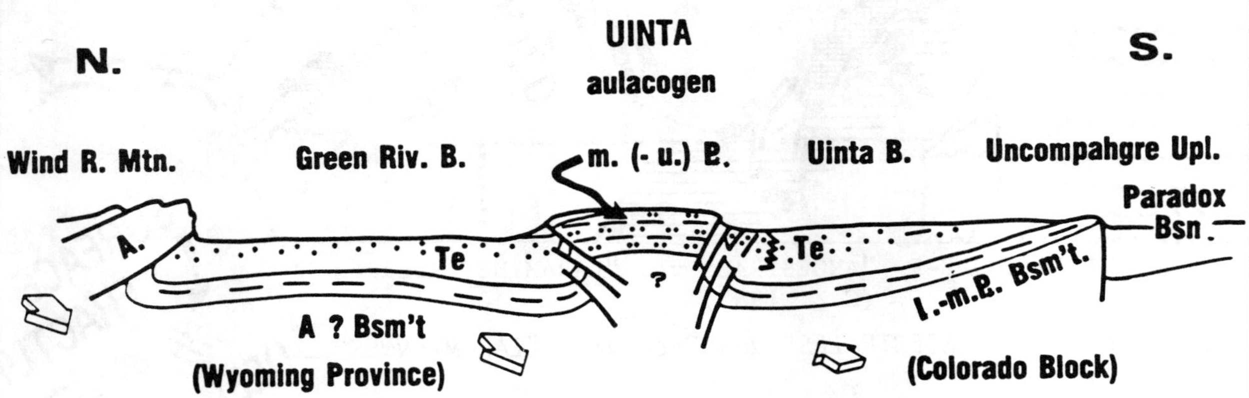

Fig. VI-19/C-D-E-F "ANCESTRAL" vs. "LARAMIDE" Rockies (G.Z. 1985)

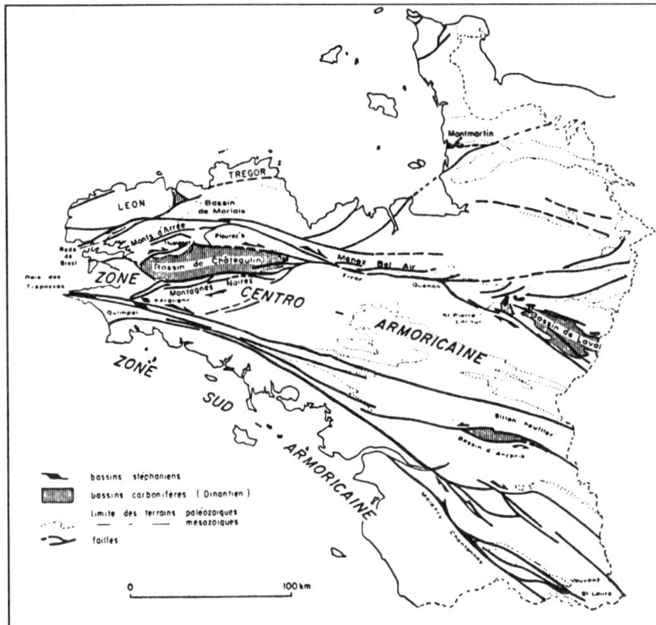

Fig. 1. — Les bassins carbonifères du Massif Armoricain.

Fig. 1. — The armorican carboniferous basins.

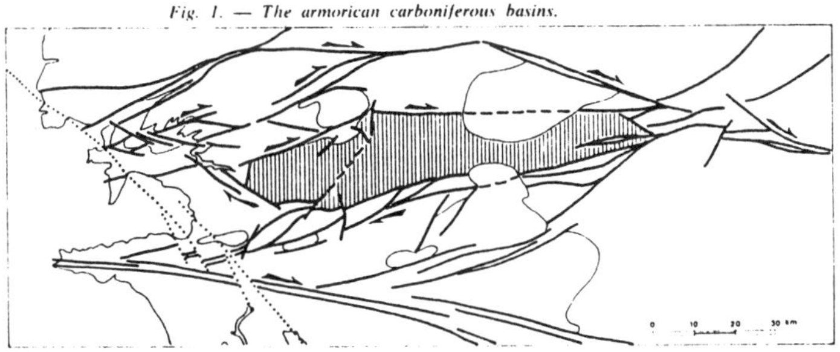

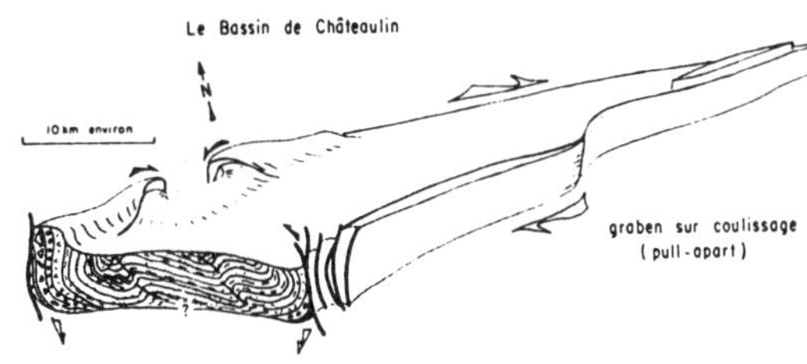

Fig. 2. — Le Bassin de Châteaulin.

Fig. 2. — The Châteaulin basin.

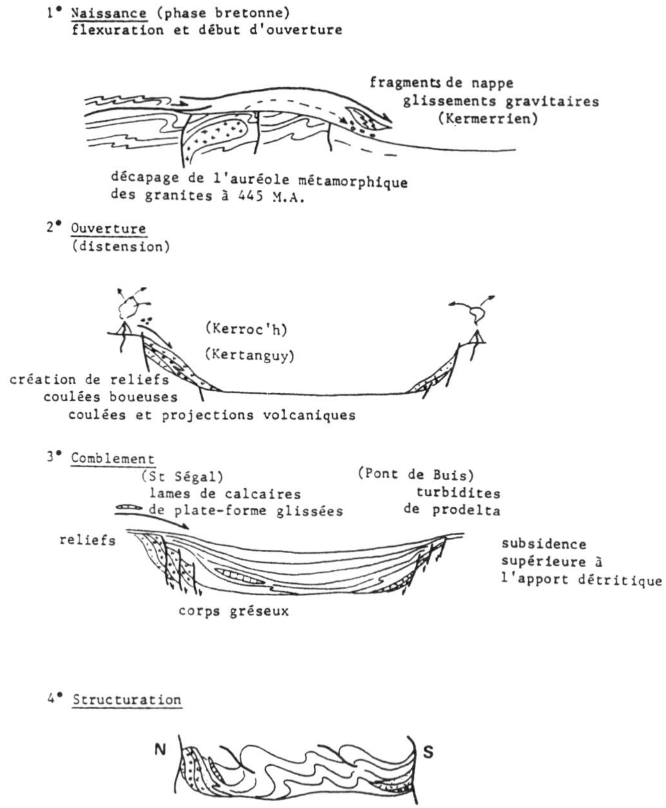

Fig. 3. — Différents stades de la vie du Bassin de Châteaulin.

Fig. 3. — Different stages of the Châteaulin basin evolution.

Evolution of the CHÂTEAULIN Basin, during the Carboniferous

Evolution du Bassin de CHATEAULIN, pendant le Carbonifère.

(de from: J. ROLET - 1984.)

Fig. VI-20 - Britanny (W. France)

Fig. VI-20 - Bretagne (France)

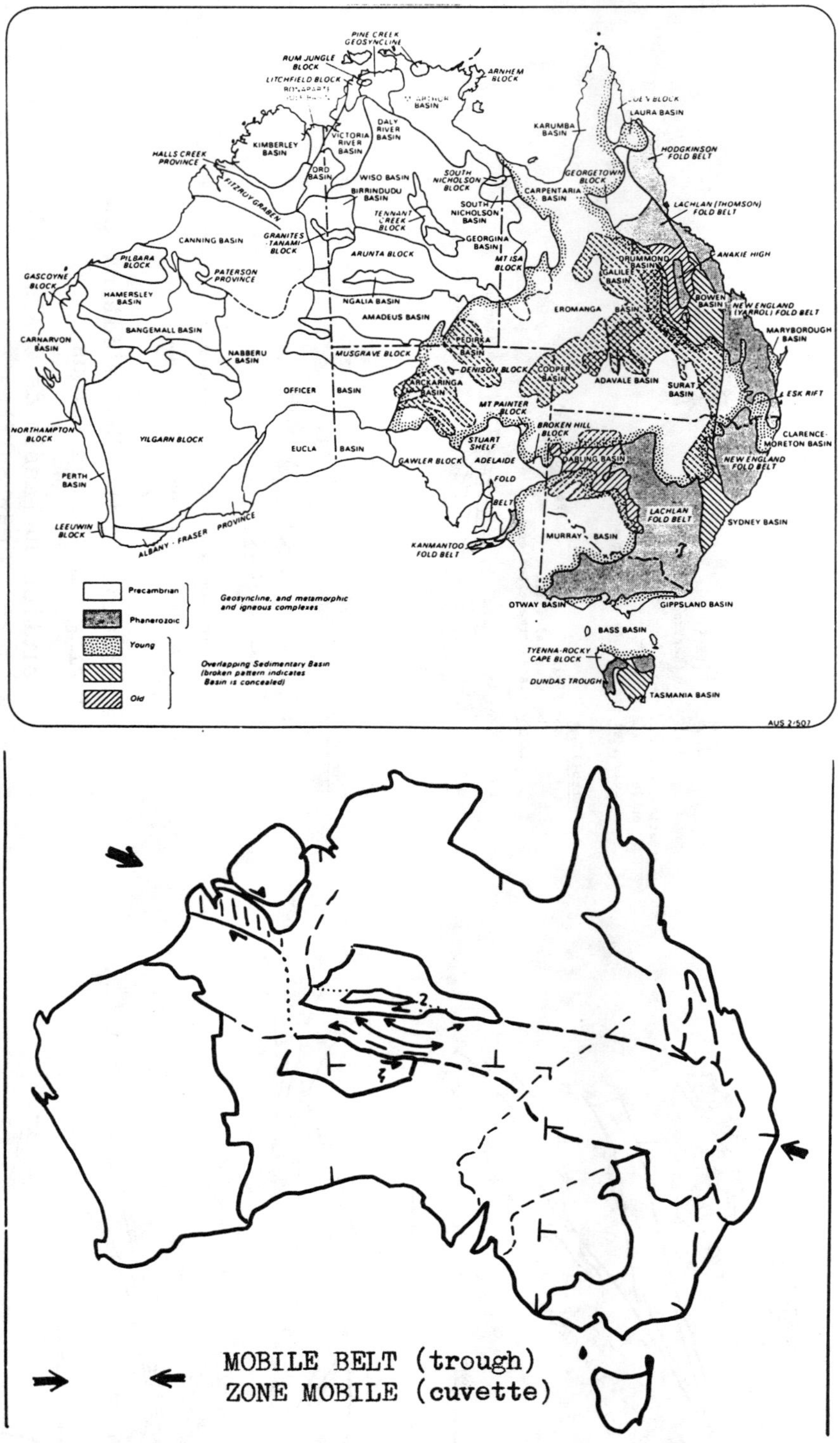

Fig. VI-21 - The Central Australian Trough (Fitzroy - Amadeus)
(G.Z.)

Fig. VI-21 - L'effondrement central d'Australie
(G.Z.)

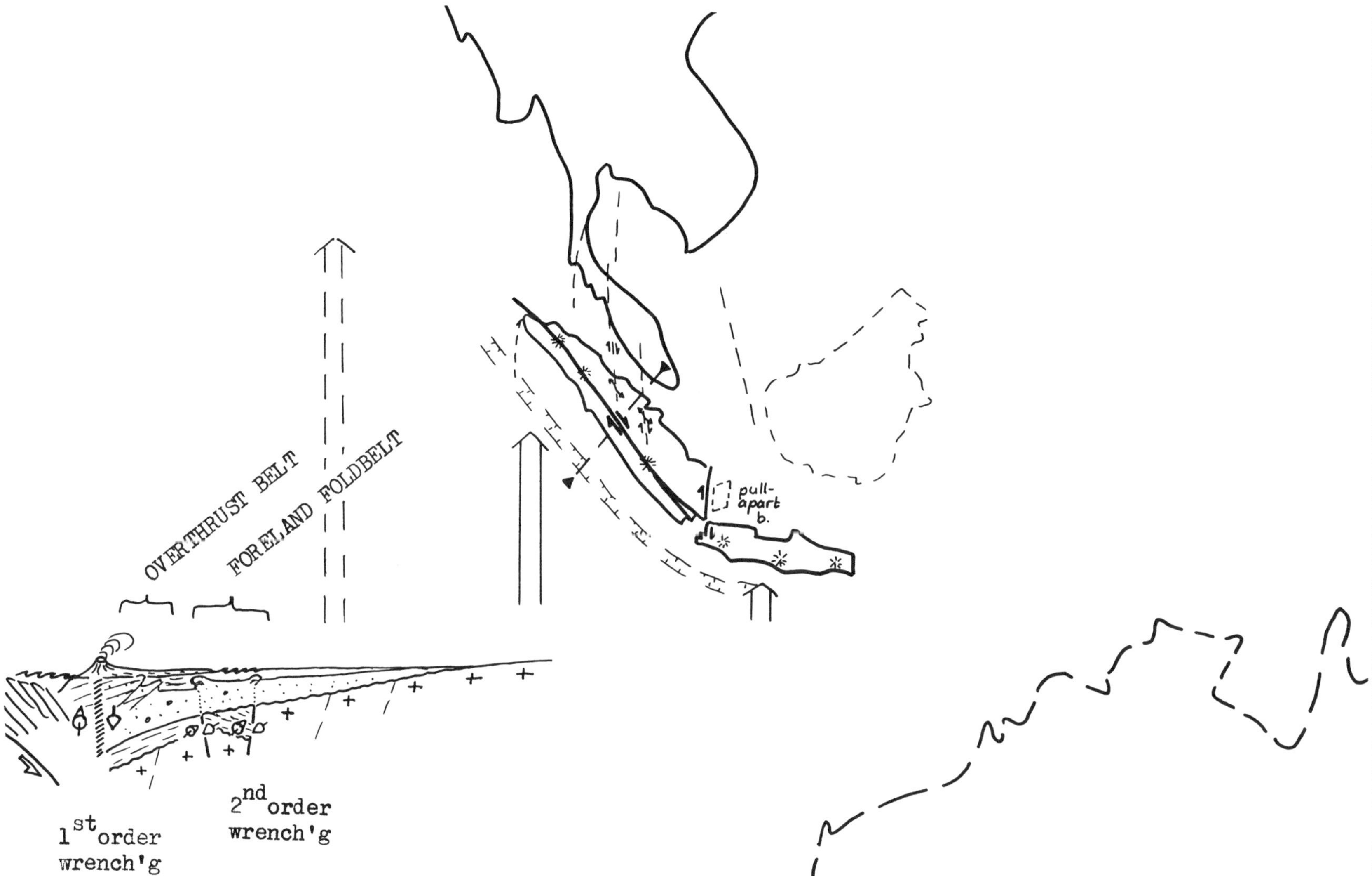

Fig. VI-22 - Western Indonesia - Malaysia (G.Z.)

Fig. VI-22 - Indonésie - Malaisie (G.Z.)

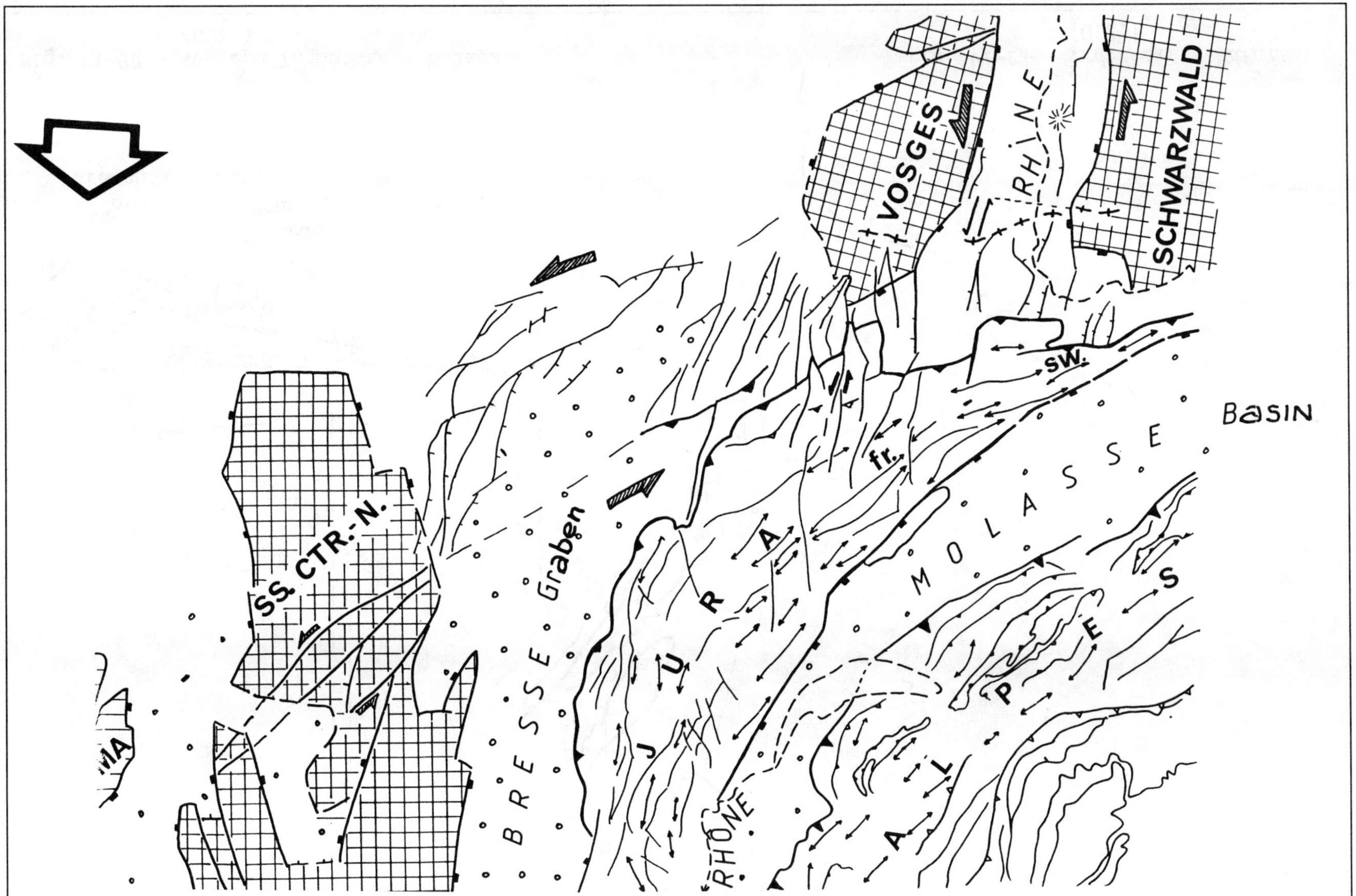

Fig. VI-23 - Eastern France (Burgundy, Jura, Rhine Rhône areas)
(RUHLAND, GELARD, BERGERAT, G.Z.)

Fig. VI-23 - L'Est de la France : Bourgogne, Jura, Région Rhin-Rhône
(RUHLAND, GELARD, BERGERAT, G.Z.)

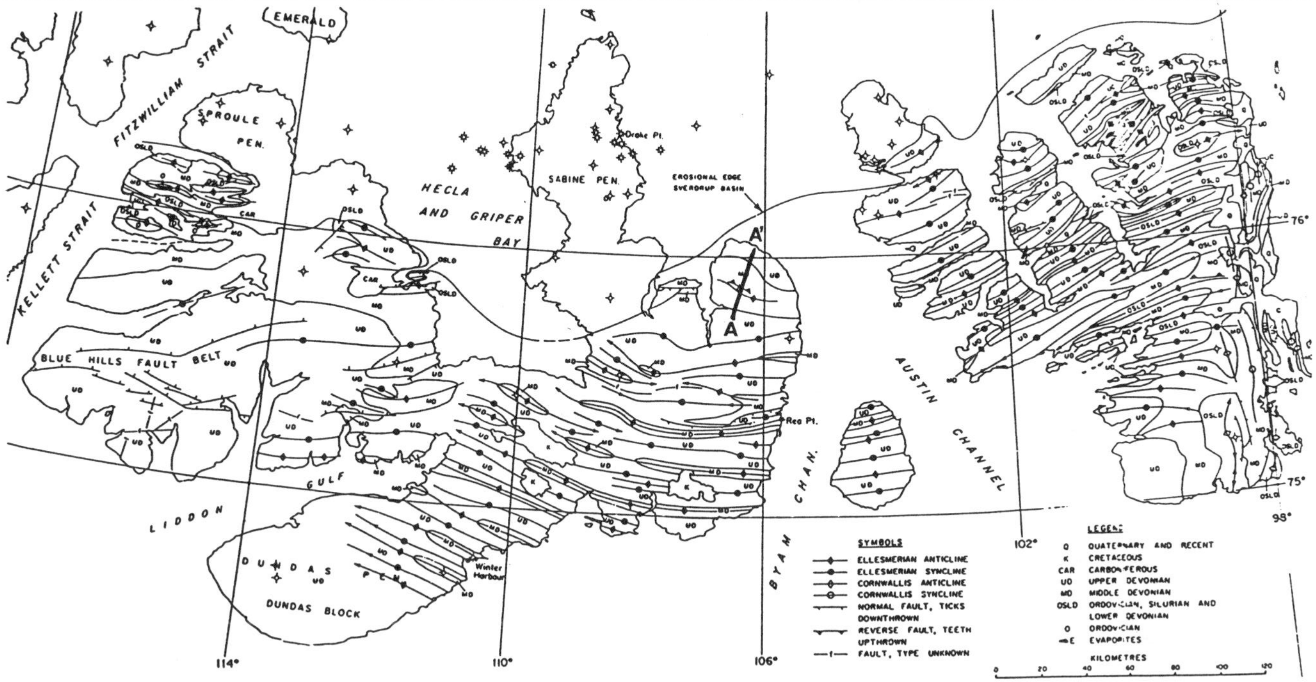

Geology of the Parry Islands Fold Belt, Bathurst and Melville Islands. After Kerr (1974), Thorsteinsson and Tozer (1964).

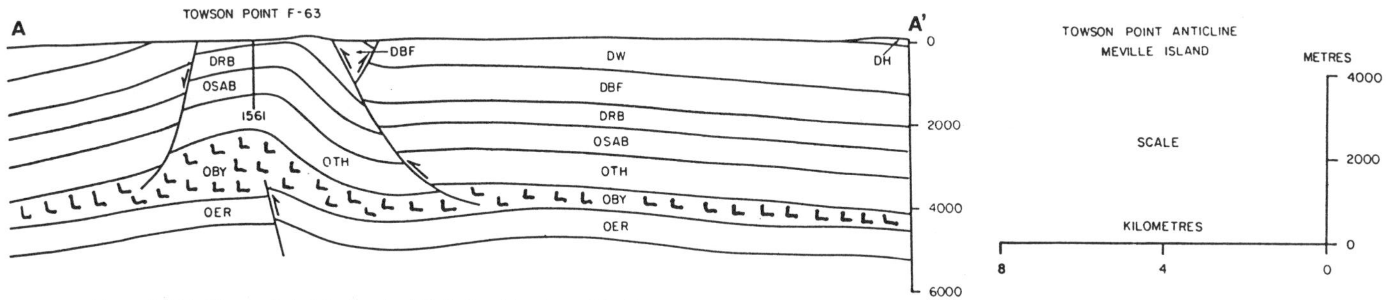

Towson Point Anticline, geological structure, line AA'. Stratigraphic units are. DG - Griper Bay, DH - Hecla Bay, DW - Weatherall, DBF - Blue Fiord, DRB - Read Bay, OSAB - Allen Bay, OTH - Thumb Mountain, OBY - Bay Fiord, OER - Eleanor River. Weatherall includes Cape de Bray.

Fig. VI-24 - Diapirism - halokinesis (Franklinian furrow, Arctic Isl.)

- *Diapirisme - Halocinèse (Sillon Franklinien, Canada)* (FOX)

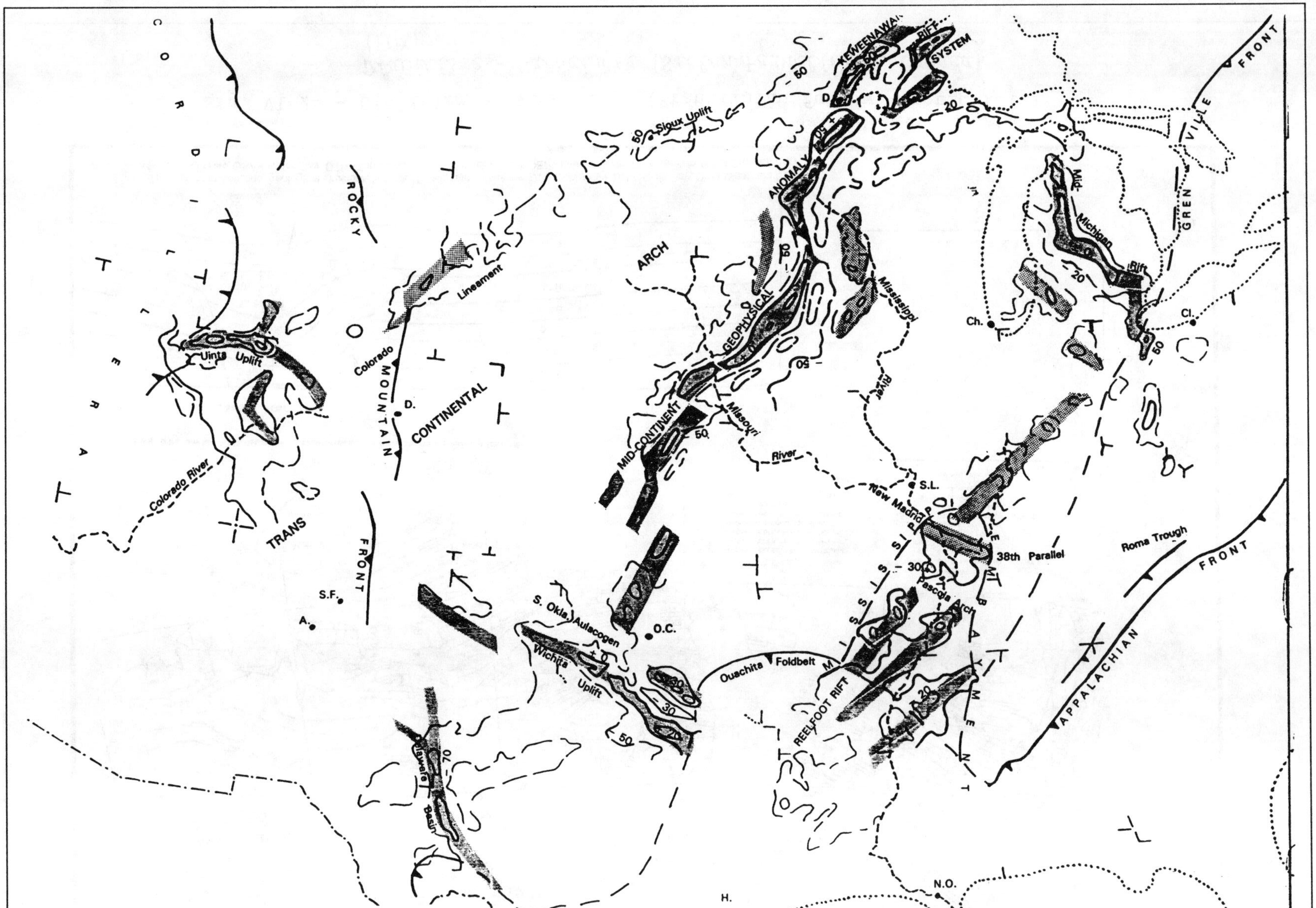

In grey : Positive Bouguer Anomalies

En gris : Anomalies positives "Bouguer"

Fig. VI-25 - Gravity image (U.S.A.)

- Image gravimétrique (U.S.A.)
(G.Z.)

FIGURES

VII. HYDROCARBON GENERATION AND HABITAT IN RELATION TO WRENCH FAULTS

HYDROCARBURES—INFLUENCE DES DECHROCHEMENTS SUR LEUR GENERATION ET HABITAT

VII-B) FIGURES

VII-B) FIGURES

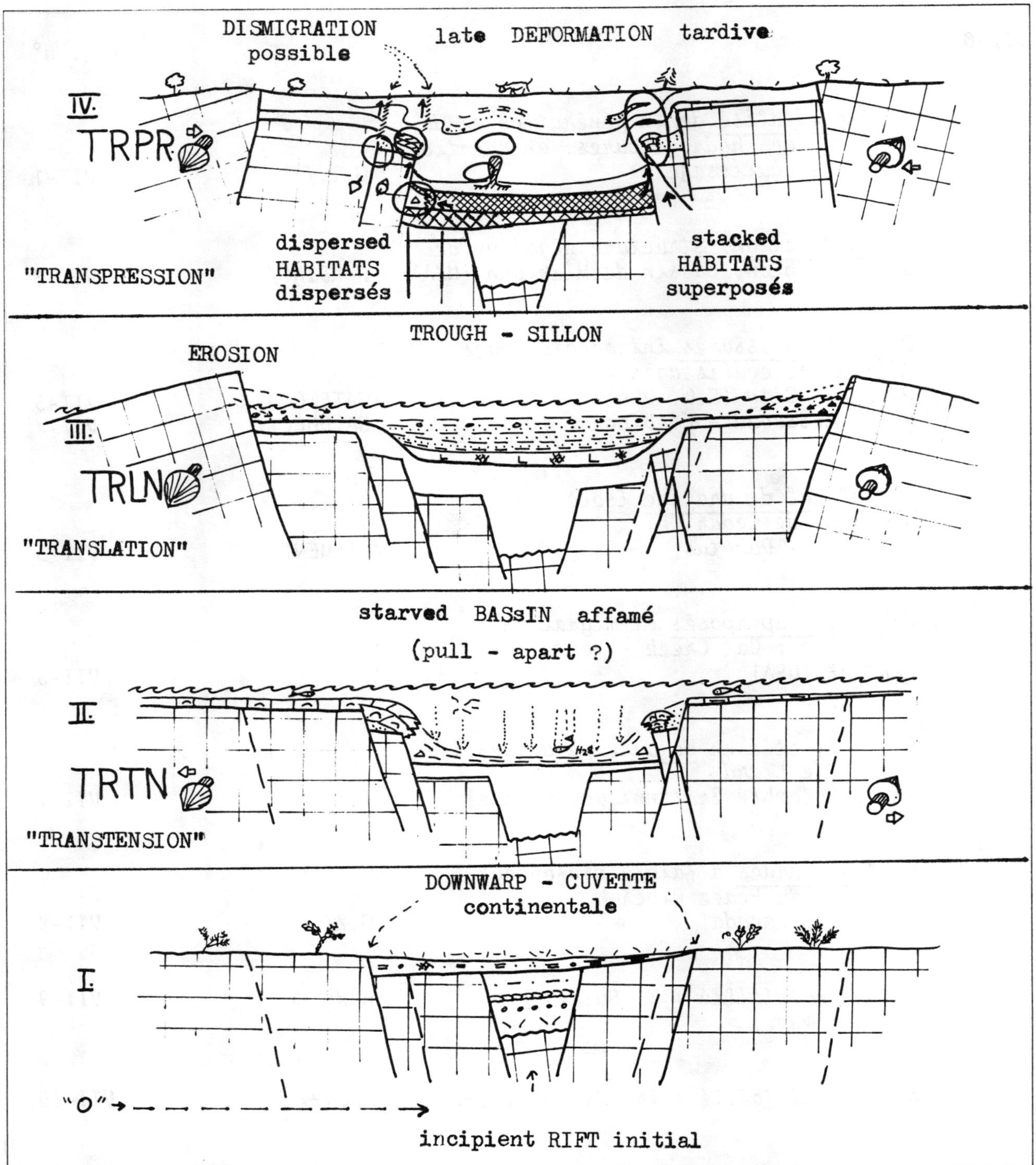

Fig. VII-1 Ideal Scheme of Generation, Trapping and possible Dismigration of Hydrocarbons, in relation to wrench-movements.

Fig. VII-1 Schéma idéalisé de la génération du piégeage et de l'éventuelle dismigration des hydrocarbures, en relation avec des coulissages.

Fig. VII-2 - Dolomitization - Scipio-Albion trend, Michigan Basin (USA)
(HARDING)

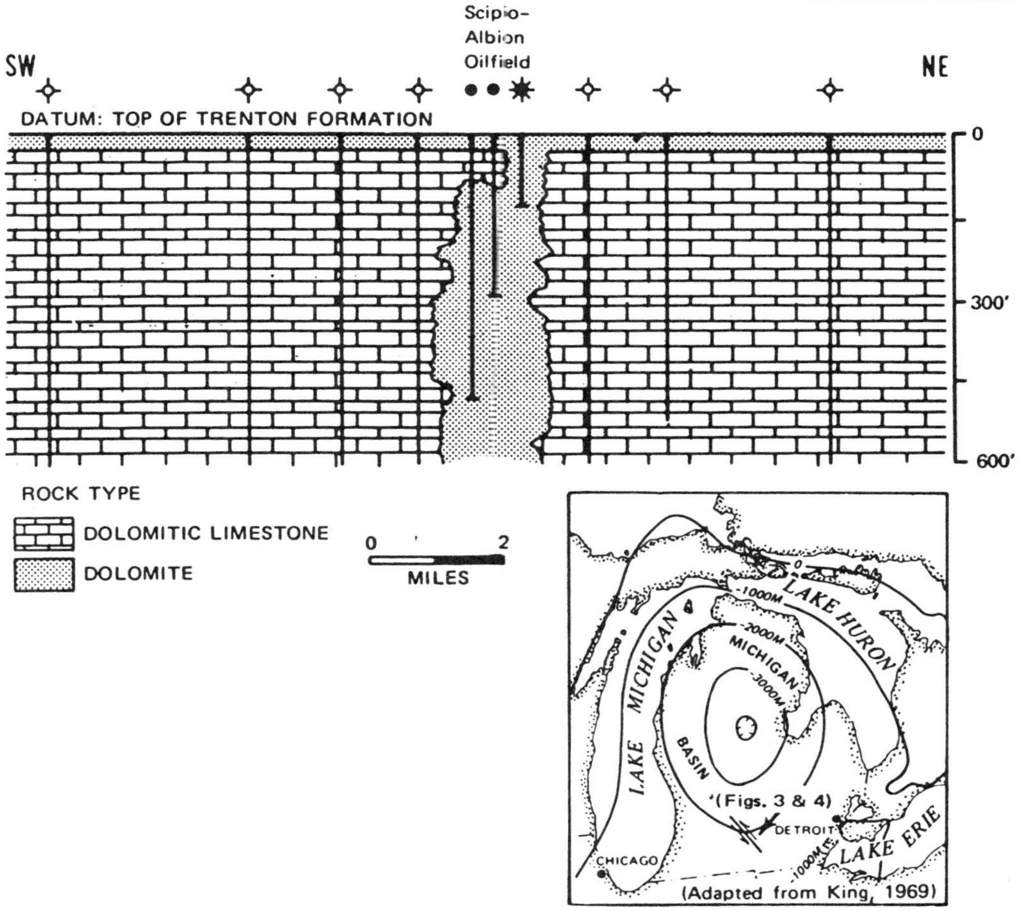

—Cross section across Scipio-Albion trend showing distribution of carbonate rock types in productive Middle Ordovician Trenton–Black River sequence (courtesy of J. N. Bubb) and area-index map contoured on top of Precambrian basement.

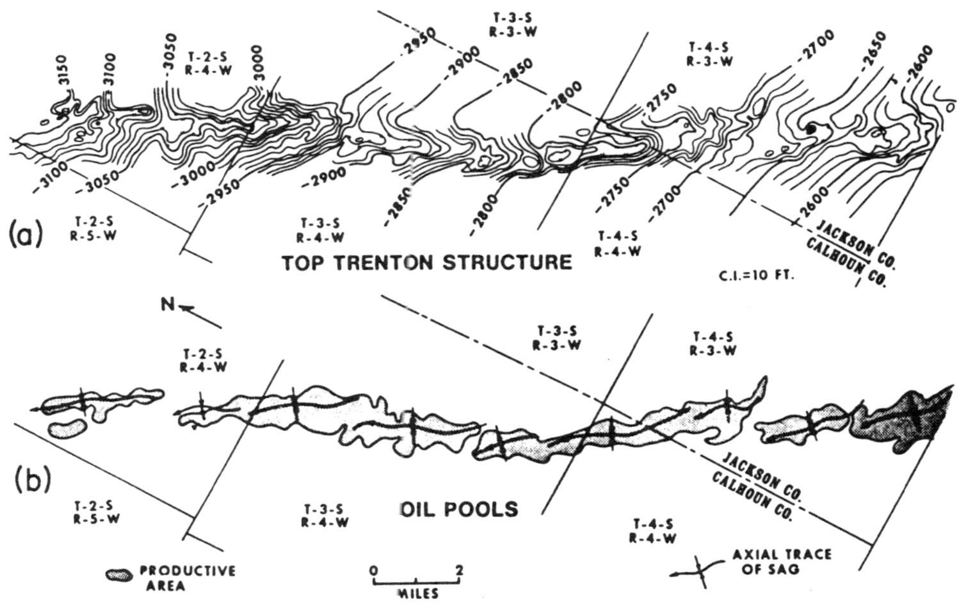

—(a) Structure-contour map of central and northern parts of Scipio-Albion trend contoured on top of Trenton formation which directly overlies productive interval (see Fig. 3). (b) Plot of main axial trace of sag.

Fig. VII-2 - Dolomitisation, - "Structure" de Scipio-Albion,
Bassin de Michigan (USA)
(HARDING)

Oklahoma City Field—Anatomy of a Giant

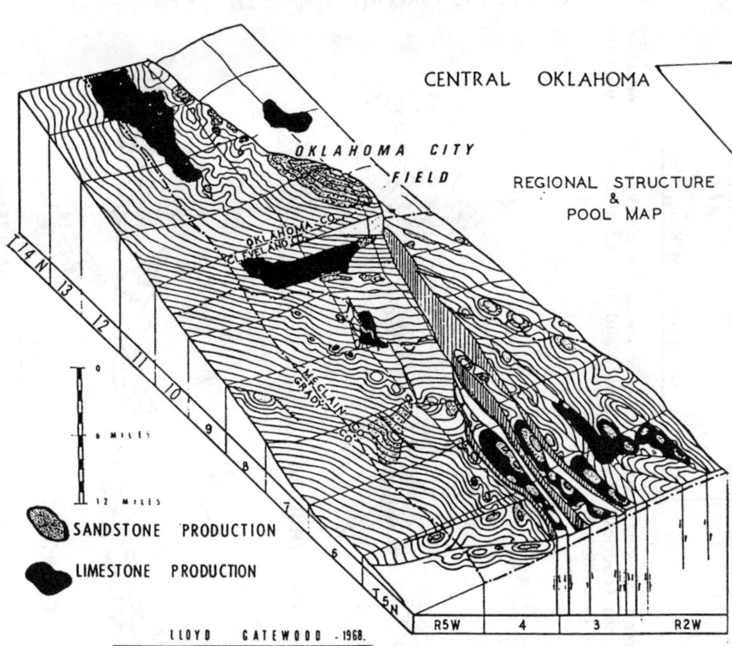

—Regional structure and pool map showing prominence of Oklahoma City uplift, which also provided environment for development of large flanking Siluro-Devonian Hunton limestone stratigraphic traps indicated by darker areas.

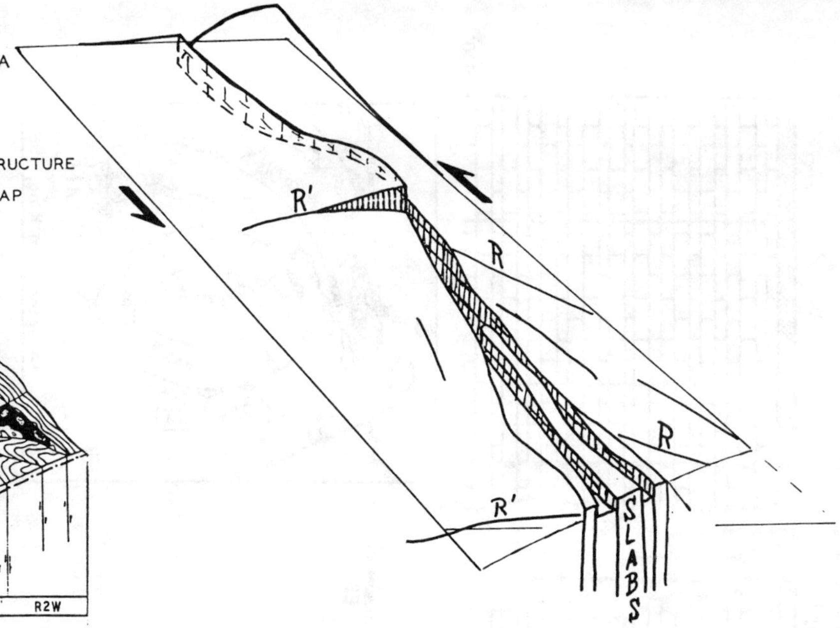

INTERPRETATION
with
WRENCHING

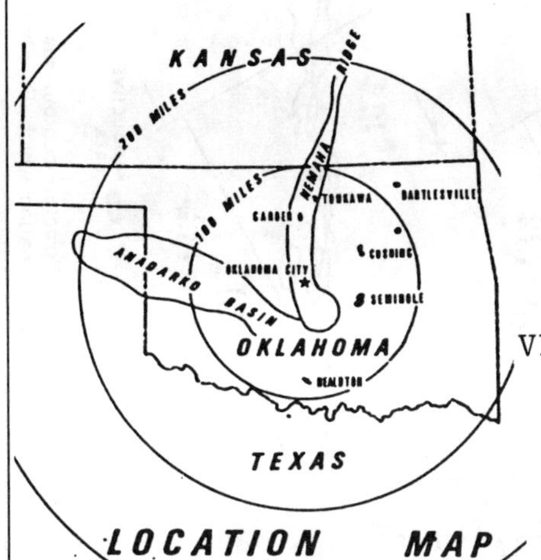

VII-3 - Large structures related to wrench-faulting :
Oklahoma City (S.W. USA)
(GATEWOOD)

- *Domes et voussures intracratoniques liés à des coulissages :*
Oklahoma City (S.W. USA)
(GATEWOOD)

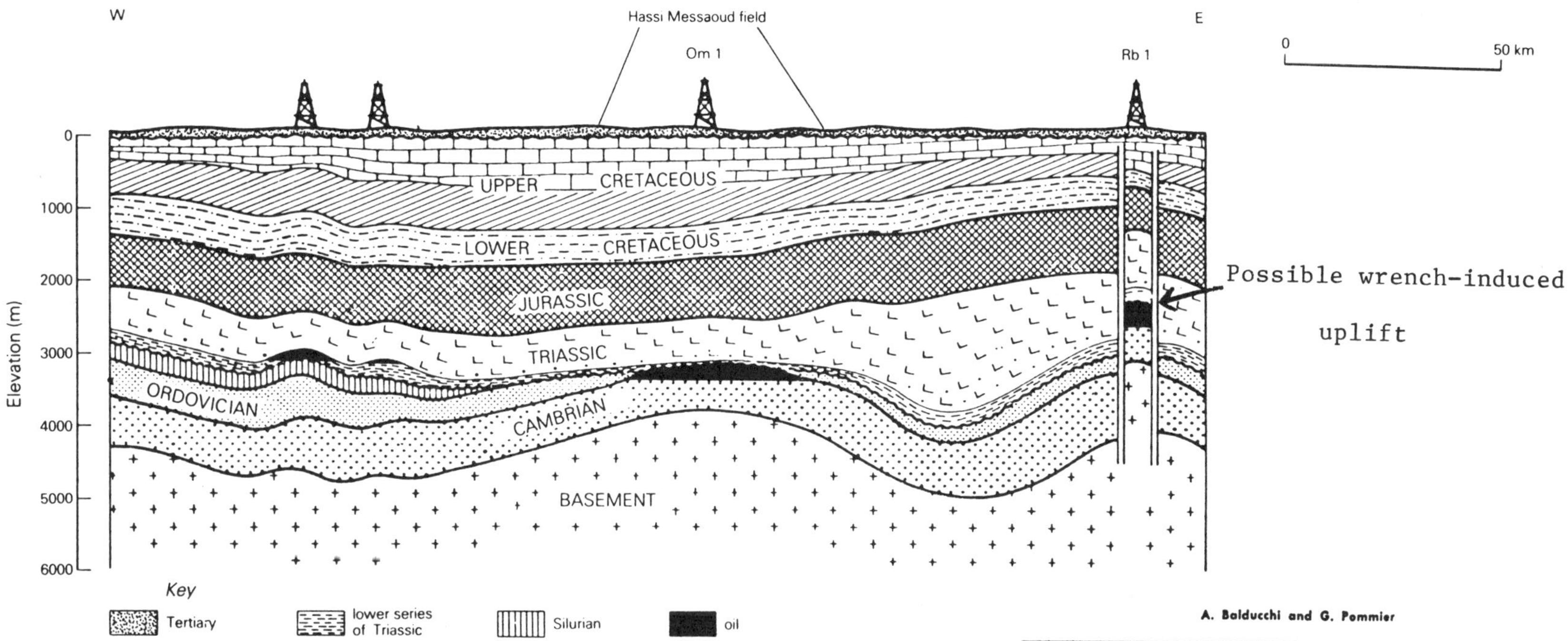

West–east cross section of the central part of the Ouargla Basin, showing the Hassi Messaoud (Om) and Rhourde el Baguel (Rb) fields. (From A. Balducchi and G. Pommier, *AAPG Memoir* 14, 1970.)

A. Balducchi and G. Pommier

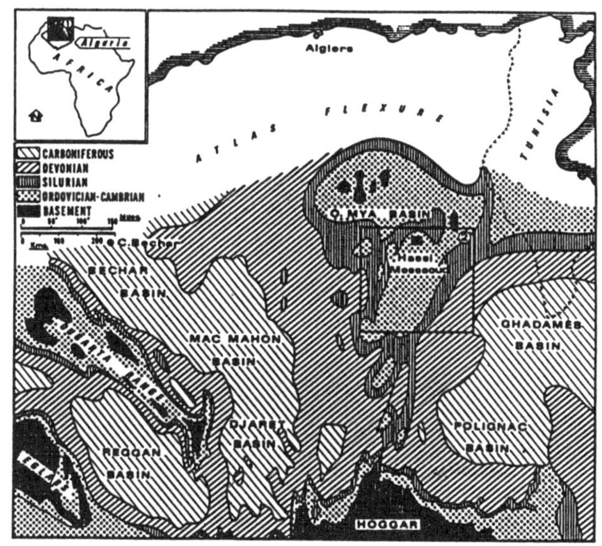

FIG. 1.—Outcrop and subcrop map of Paleozoic in northern Sahara (Algeria and Tunisia). Shows location of Hassi Messaoud. Outlined rectangle is area of Figure 2.

Fig. VII-4 - <u>Large structures</u> related to wrench-faulting :
Hassi Messaoud & El Biod (Sahara)
(BALDUCCI)

- *Domes et voussures intracratoniques liés à des coulissages :*
Hassi Messaoud - El Biod
(BALDUCCI)

CLEMENT (1987)

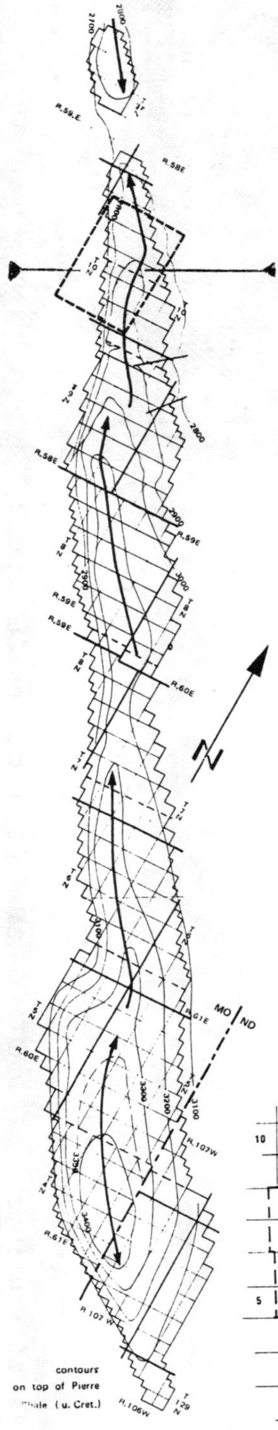

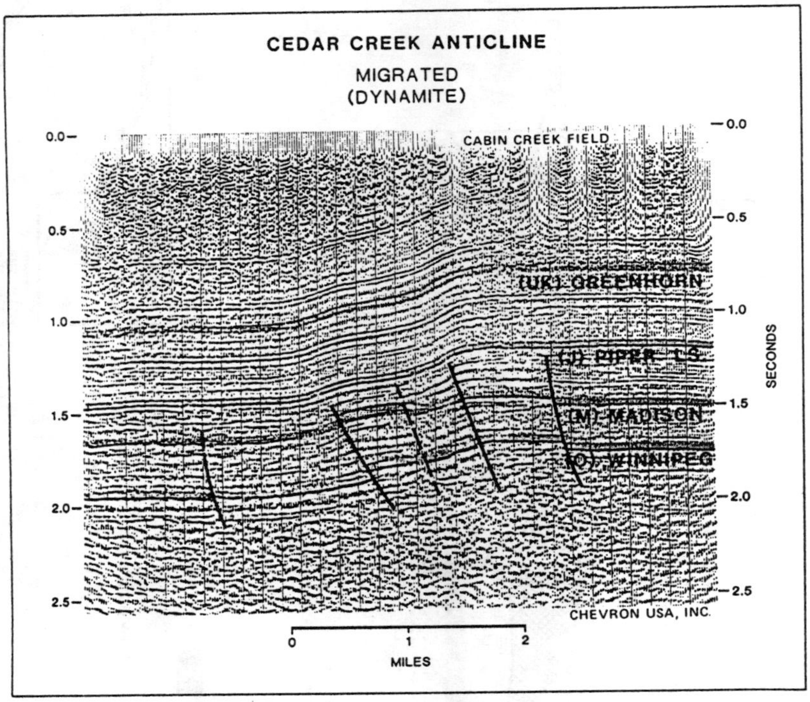

—Seismic line across Cabin Creek field.

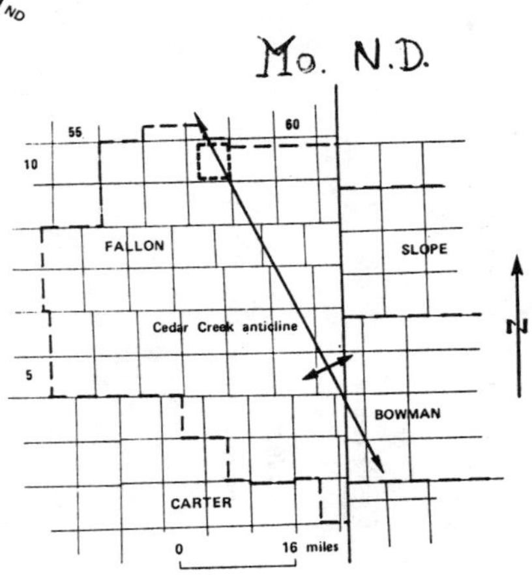

Fig. VII-5 - Isolated drape-anticline : Cedar Creek field
(Montana N. & S. Dakota)

- *Anticlinal de drapage isolé : gisement de Cedar Ck.*
(Montana - Dakotas)

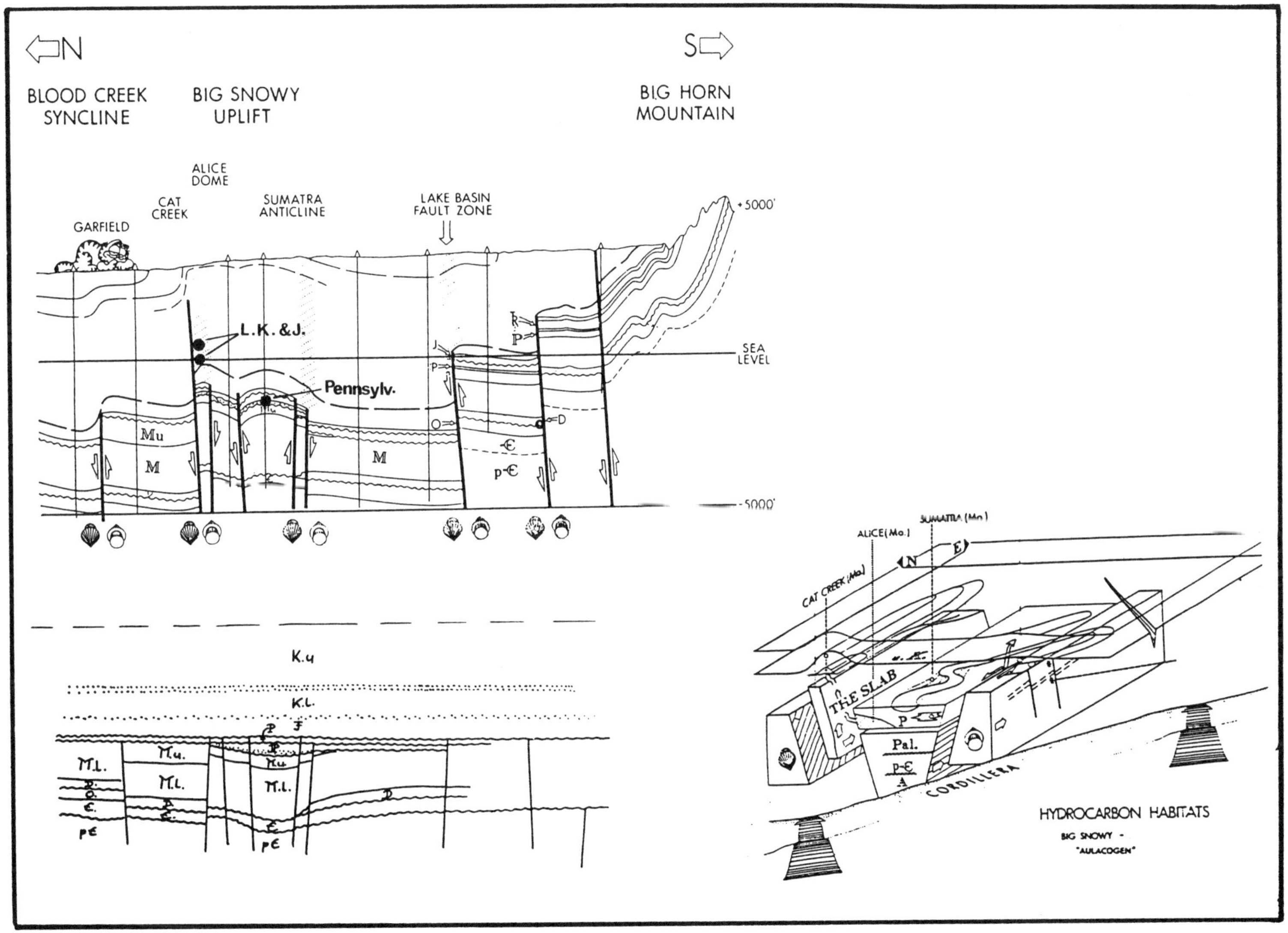

Fig. VII-6 - Habitats superposés et migration verticale ; Cat Creek - Montana Central (USA) (G.Z.)

- Stacked habitats, vertical migration - Cat Creek - Central Montana area (USA)

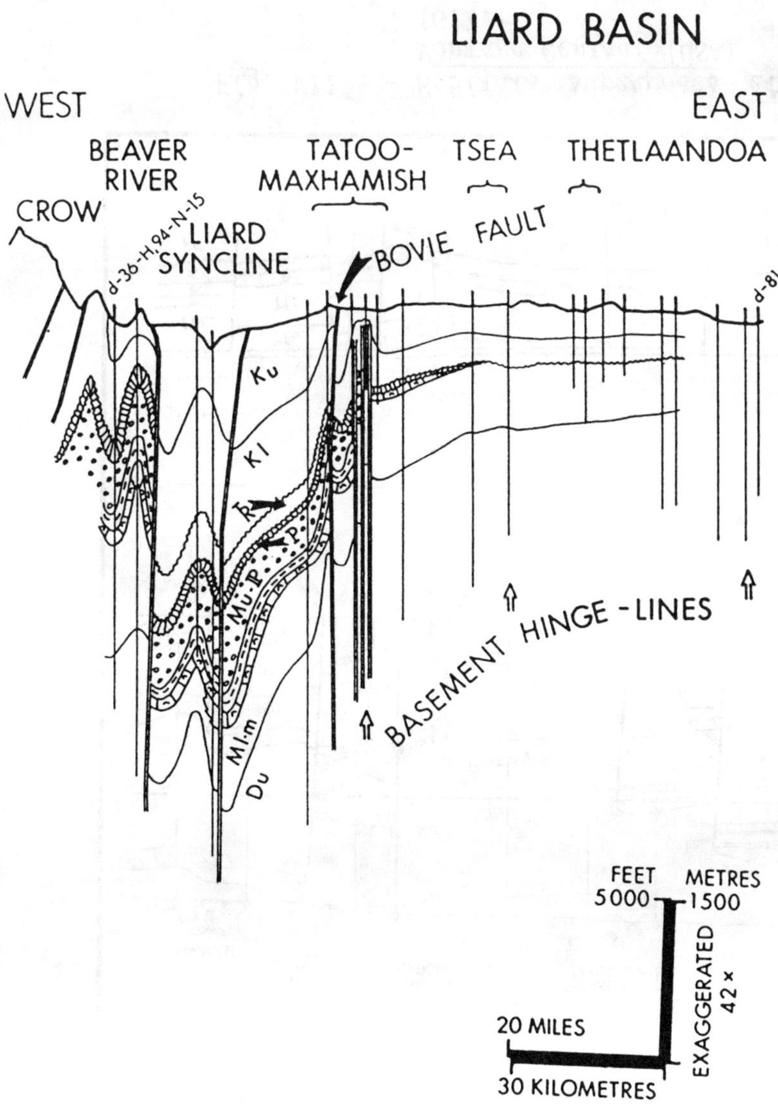

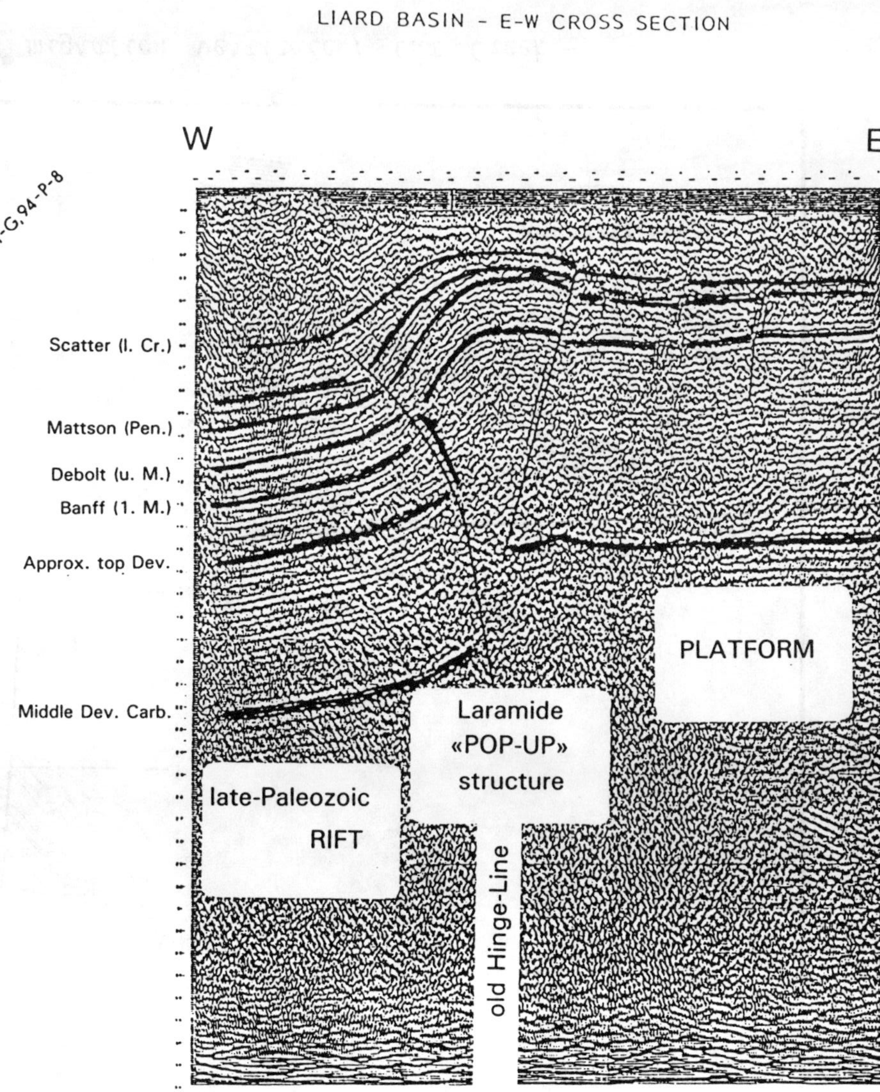

Fig. VII-7 - Graben-edge blocks : Liard Basin (N.W. Canada) (G.Z.)

- *Blocs éjectés en bordure de graben : Bassin de Liard (n.e. Colombie Britannique, Canada)*

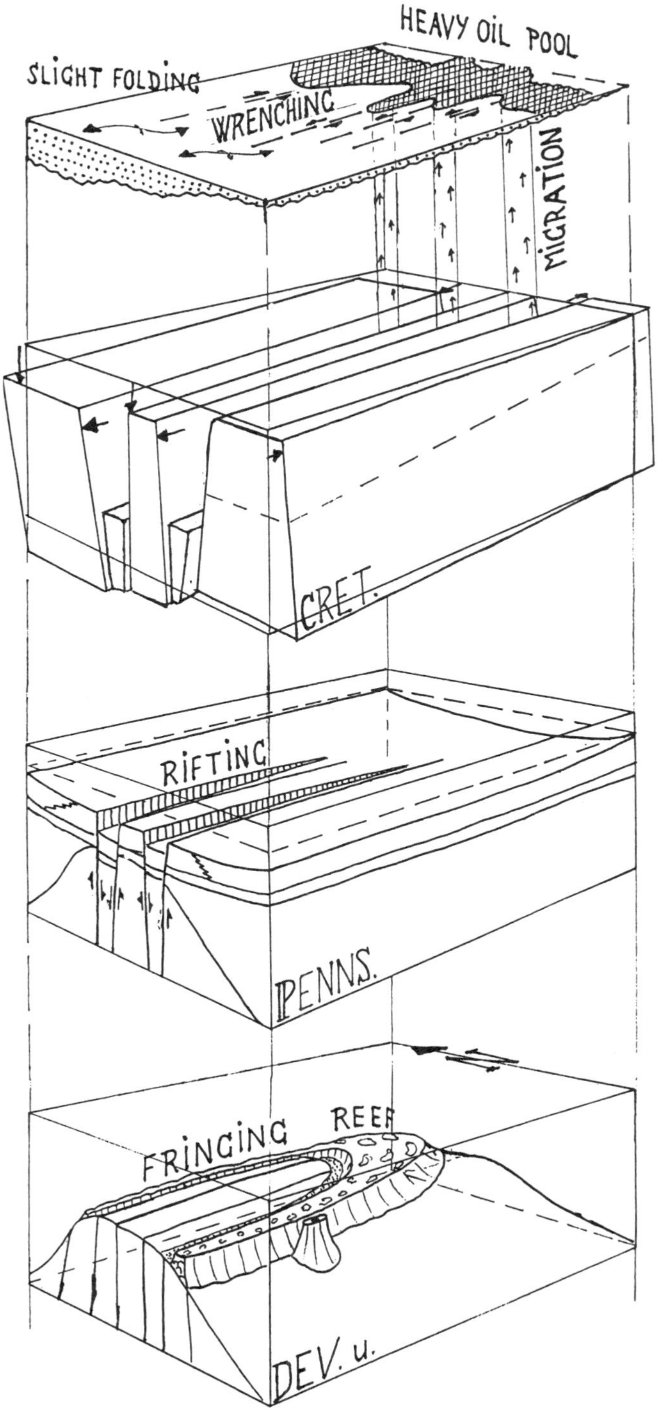

slight
WRENCHING
&
westerly TILTING

weak
EXTENSION

± steady
UPLIFTING

Fig. VII-8 - Shallow heavy-oil habitat : Peace River oil sands (Alberta, W. Canada)

- *Huiles lourdes à faible profondeur : gisement de Peace River (Alberta, Canada)*
(G.Z)

From/de: HARDING -1974

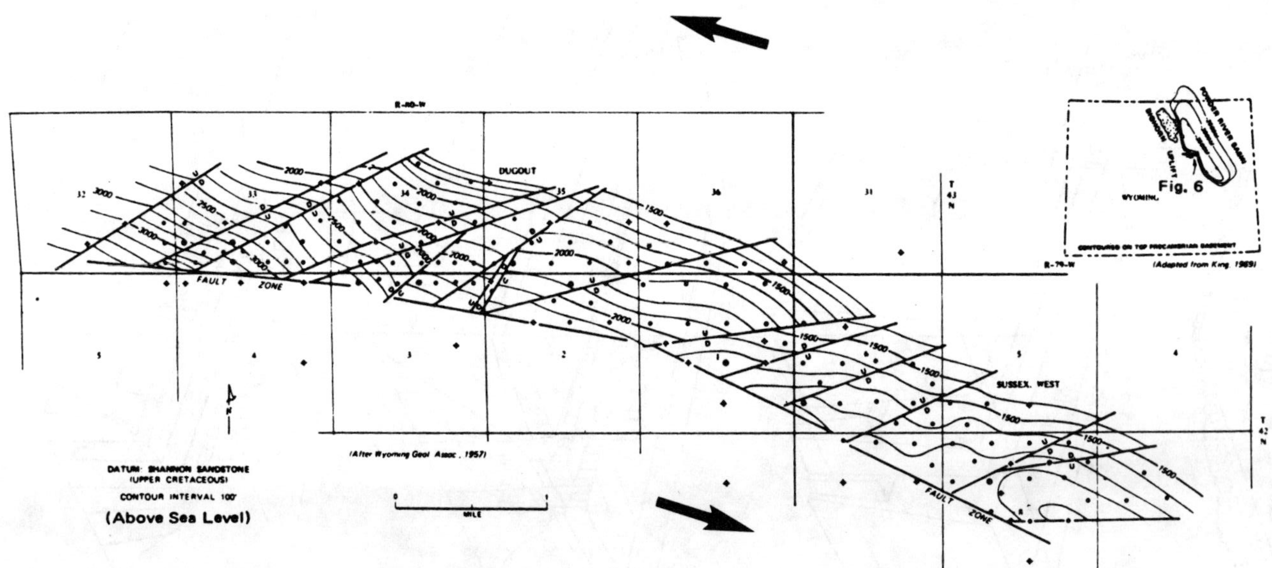

—Structure of West Sussex–Dugout oil field, Powder River basin, Wyoming.

Fig. VII-9 - Faulted anticline : W. Sussex-Dugout (Wyo.) (HARDING)

Fig. VII-9 - Anticlinal faillé : W. Sussex - Dugout (Wyo.) (HARDING)

From/de: MOODY - 1973

Petroleum Exploration Aspects of Wrench-Fault Tectonics

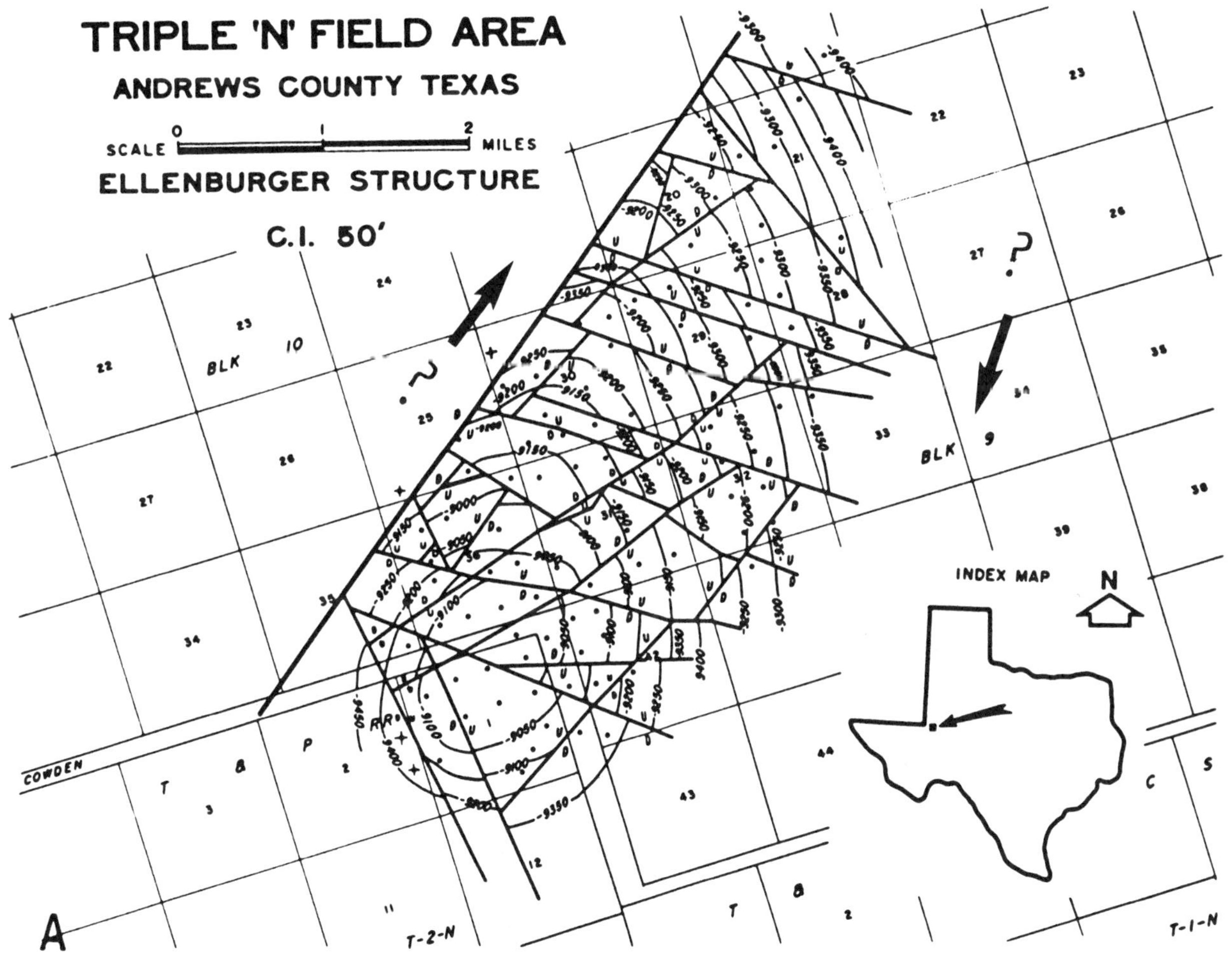

Fig. VII-10 - Faulted anticline : Triple "N" (Tex.) (MOODY)

Fig. VII-10 - Anticlinal faillé : Tripe "N" (Tex.) (MOODY)

From/de:
SALES
1971

STRUCTURE OF THE NORTHERN MARGIN OF THE GREEN RIVER BASIN, WYOMING

LOST SOLDIER AREA
SPACE RELATIONSHIP

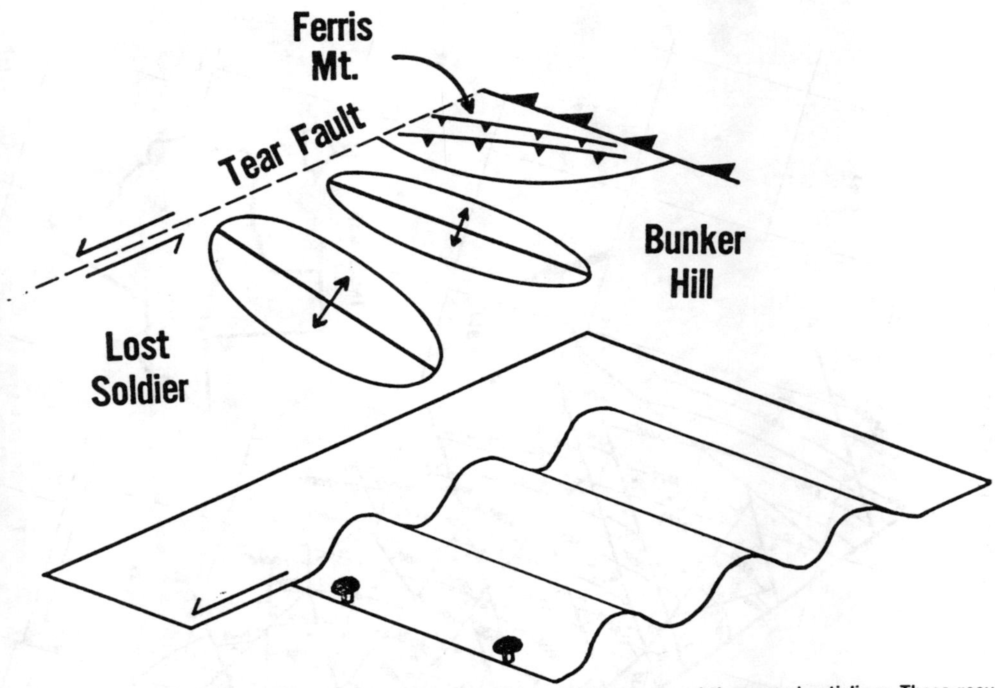

. Diagrammatic representation of the space conditions that have caused the named anticlines. These result from left-lateral distributive tearing as "South Granite Mountains Shortening" is distributed westward to allow "Wind River Shortening."

for position of "Lost Soldier" field, see Fig. VI. 17 (center)

Fig. VII-11 - Drag-fold : Lost Soldier (Wyo.)
(SALES)

Fig. VII-11 - Anticlinal d'entraînement : Lost Soldier (Wyo.)
(SALES)

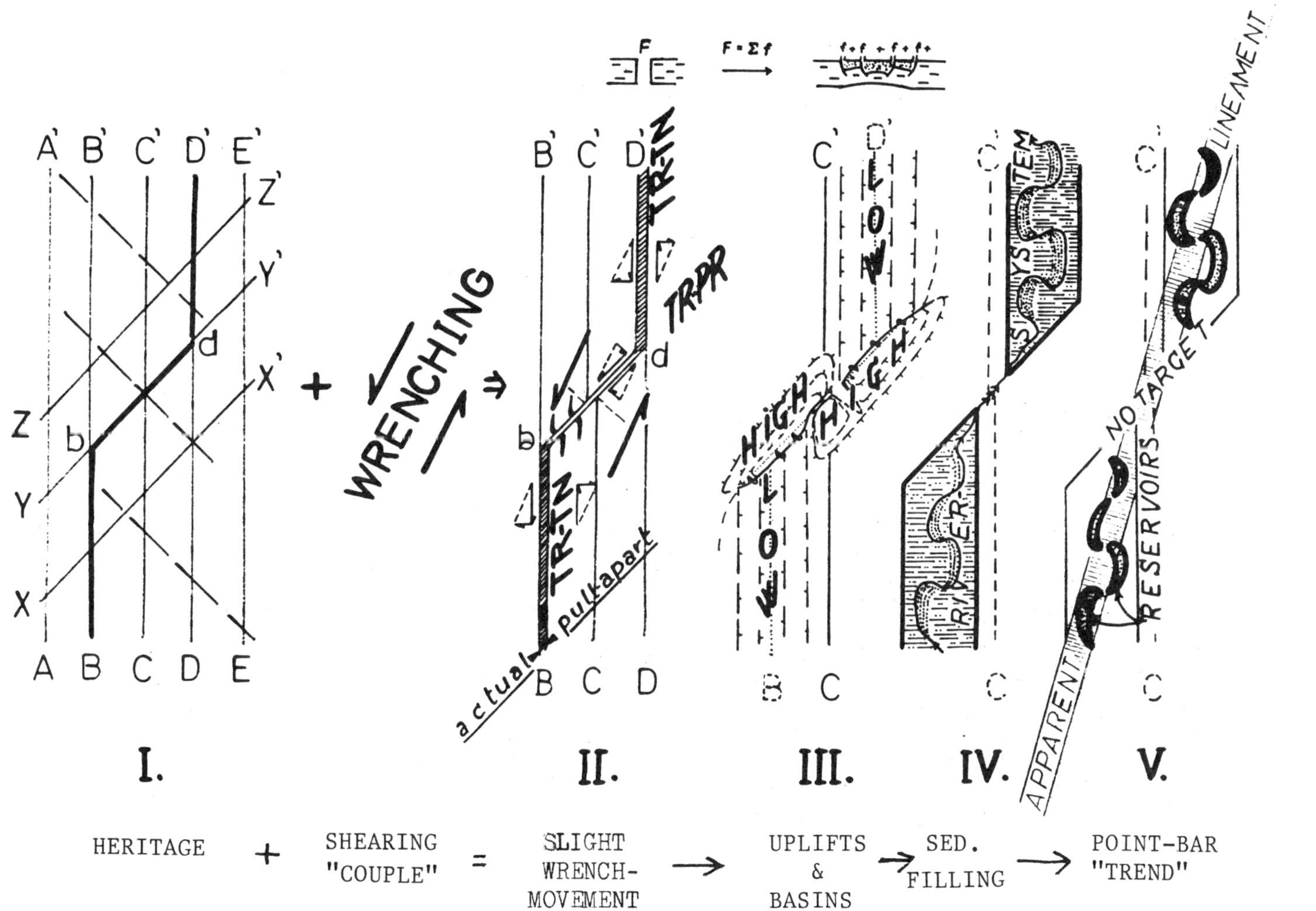

Fig. VII-12 - The elusive trend
(G.Z.)

Fig. VII-12 - Le "trend" discontinu
(G.Z.)

FIGURES

VIII. WORKSHOP-EXERCISES

EXERCICES

VIII. WORKSHOP - EXERCISES

1) "CREASE of the TROUSER"

2) FAULTS and FOLDS - experiment :

3) "WHAT AN INNOCENT FOLD CAN HIDE"

4) "Where is the MASS-BALANCE zone ?"

5) THINNING of the crust through STRETCHING

6) Flattening out of folds or "HOW MUCH IS TOO MUCH" ?

7) DRAPE FOLD OR DRAG FOLD ?

8) LOZENGE - BASIN

9) SCISSOR - and - SCOTCH-TAPE exercice

VIII. EXERCICES

1) "PLI du PANTALON"

2) FRACTURES et PLIS - Expérience :

3) "QUE PEUT CACHER UN SIMPLE PLI ?"

4) "Qu'est donc advenu au BILAN DES VOLUMES ?"

5) AMINCISSEMENT de la croûte par ETIREMENT

6) "Jusqu'où peut-on (ou ne doit-on pas) ALLER TROP LOIN ?

7) DRAPAGE OU ENTRAINEMENT ?

8) BASSIN LOSANGIQUE

9) Exercice des "CISEAUX et du RUBAN ADHESIF"

EXERCISES
EXERCICES

n° 1.

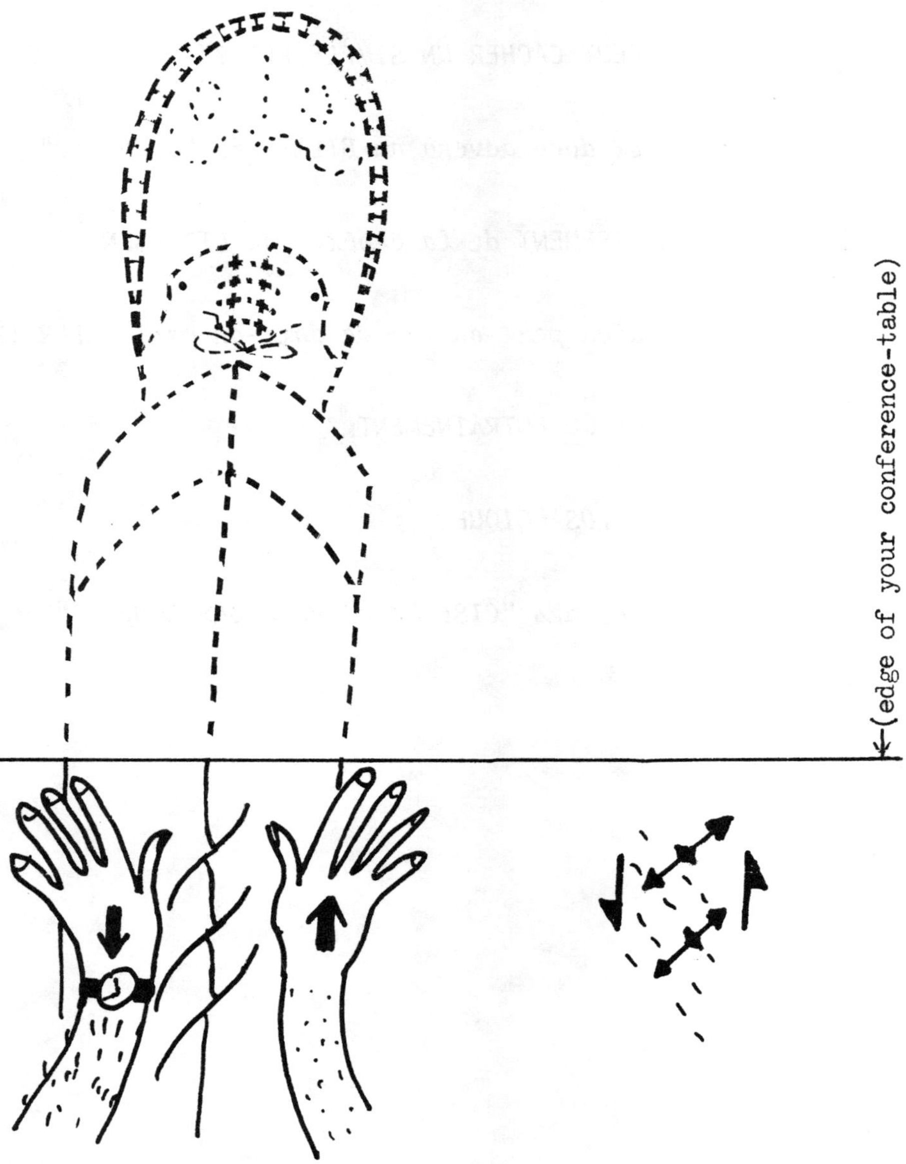

Fig. VIII-1 "Crease of the trouser" (or slacks)

"Pli du pantalon"

Fig. VIII-2/A Faults and folds in experiment "SYNTHETIC WRENCHING" : external and internal wrenching identical (Riedel's "R" fractures)

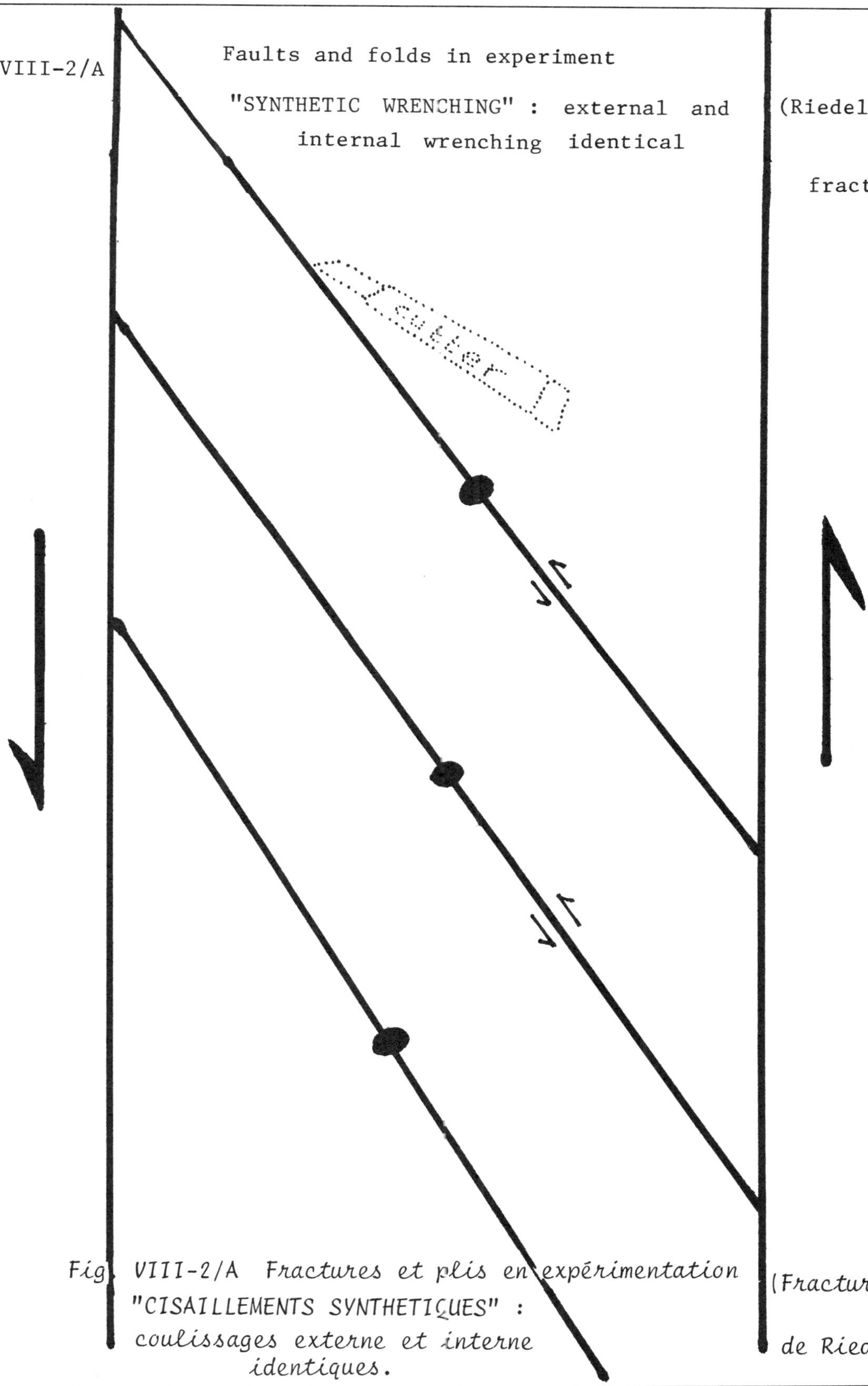

Fig. VIII-2/A *Fractures et plis en expérimentation "CISAILLEMENTS SYNTHETIQUES" : coulissages externe et interne identiques.* *(Fractures "R" de Riedel)*

Fig. VIII-2/B Faults and folds in experiment "ANTITHETIC WRENCHING" : external and internal slippage opposite

(Riedel's "R'" fractures)

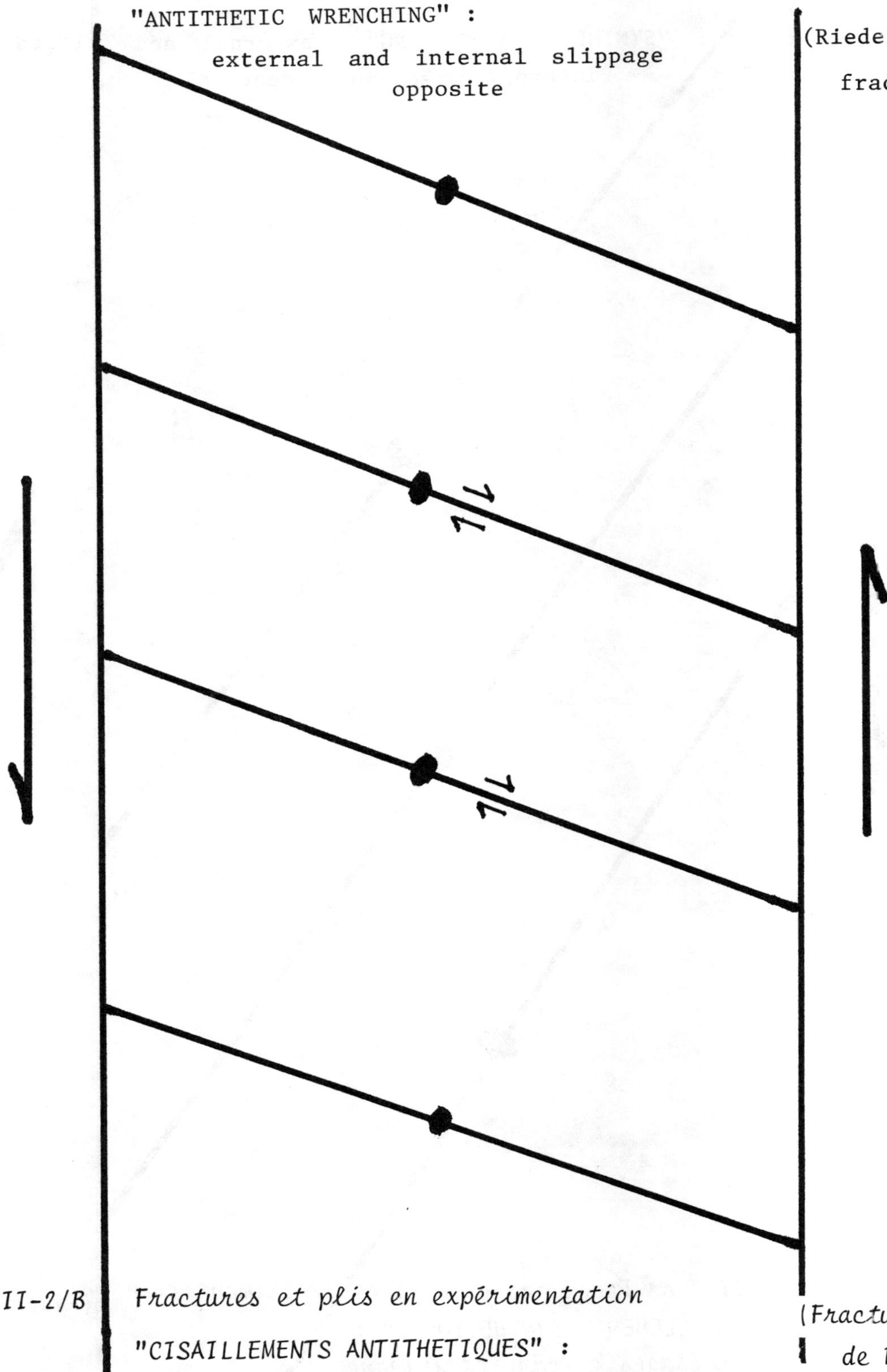

Fig. VIII-2/B *Fractures et plis en expérimentation* "CISAILLEMENTS ANTITHETIQUES" : *coulissages externe et interne en sens opposé*

(Fractures "R'" de Riedel)

Fig. VIII-3/A "What an innocent fold can hide" *Fig. VIII-3/A*

"Que peut cacher un simple pli"

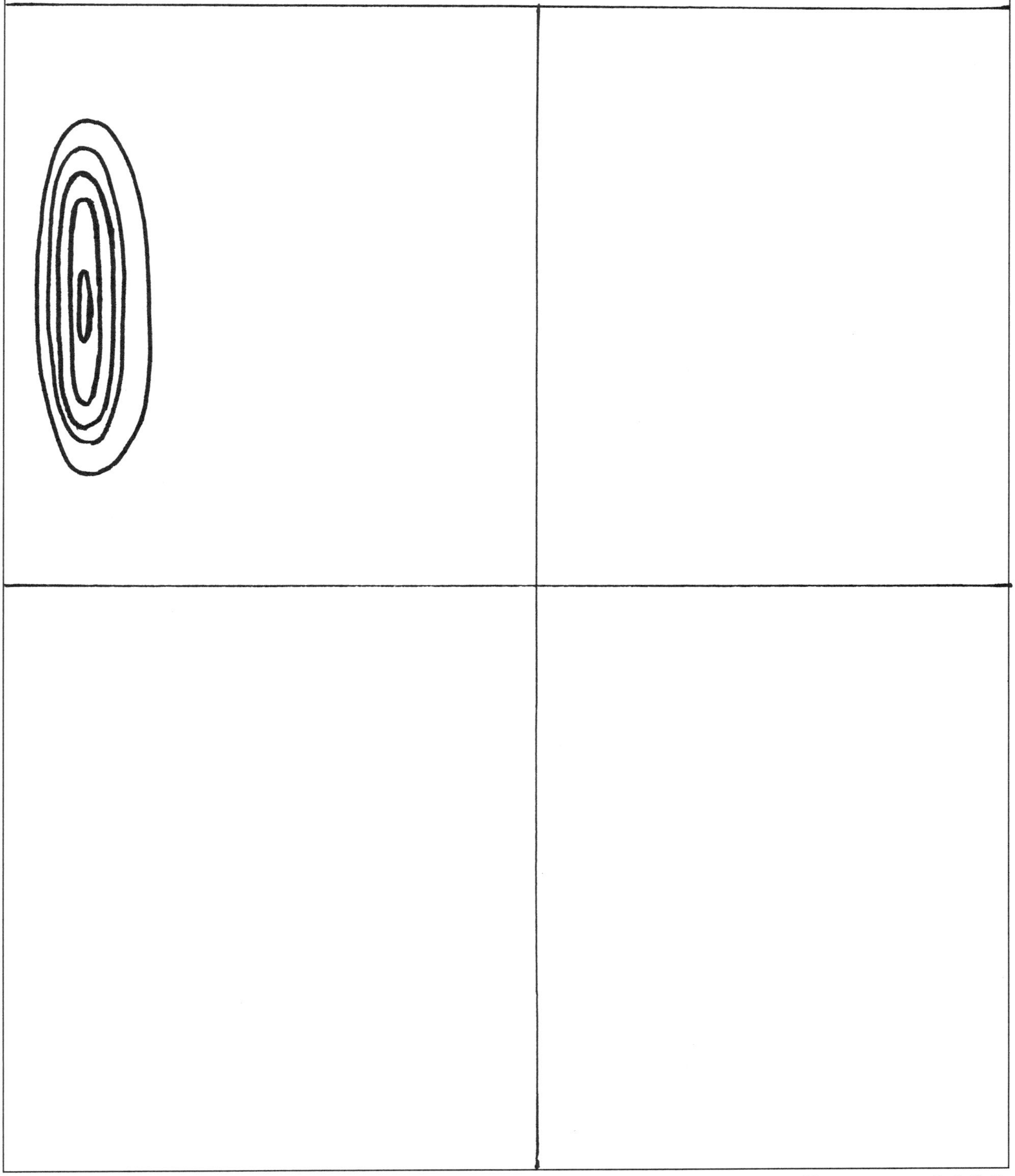

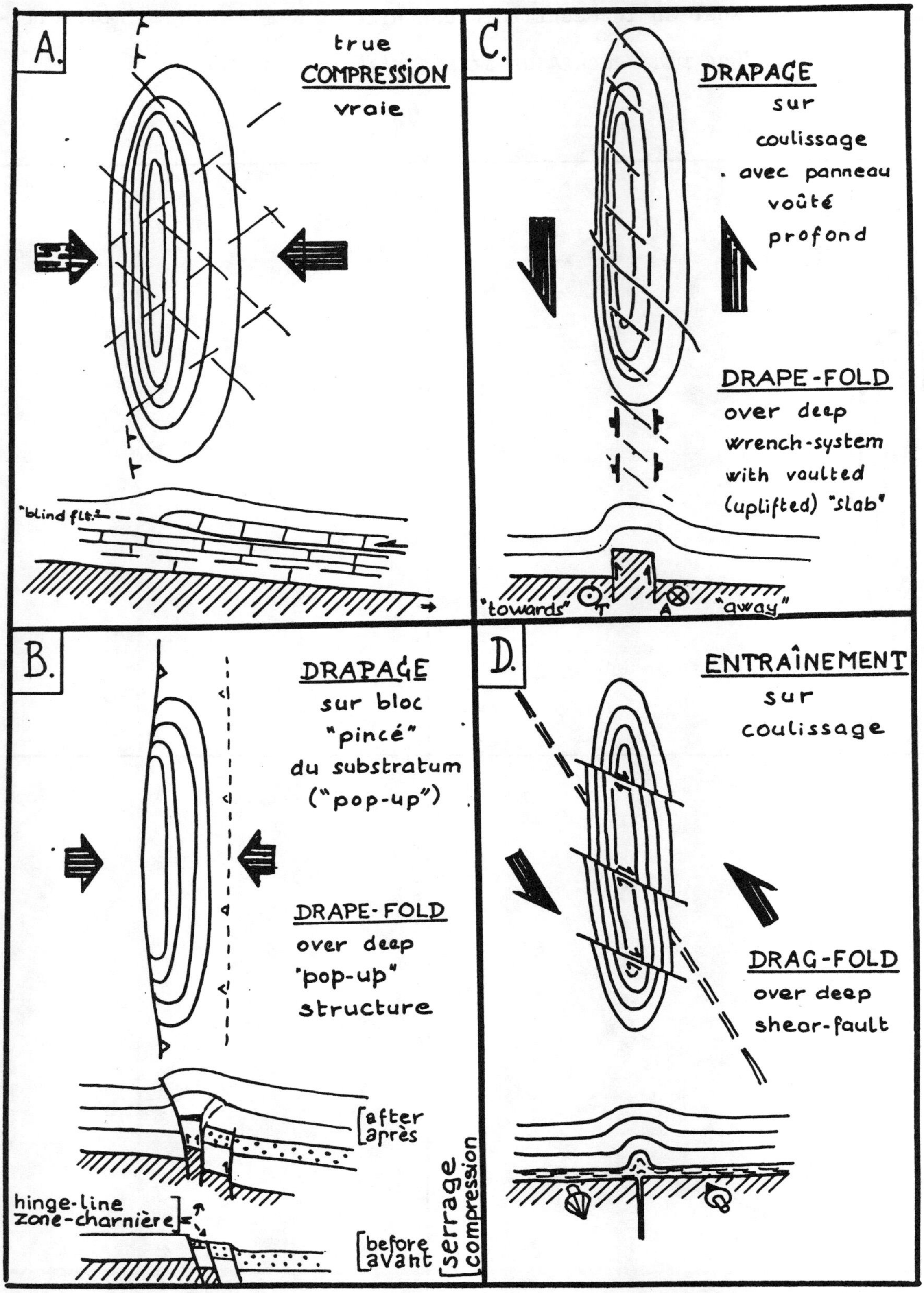

Fig. VIII-3/B Solutions

Fig. VIII-3/B Solutions

E X E R C I S E S - E X E R C I C E S

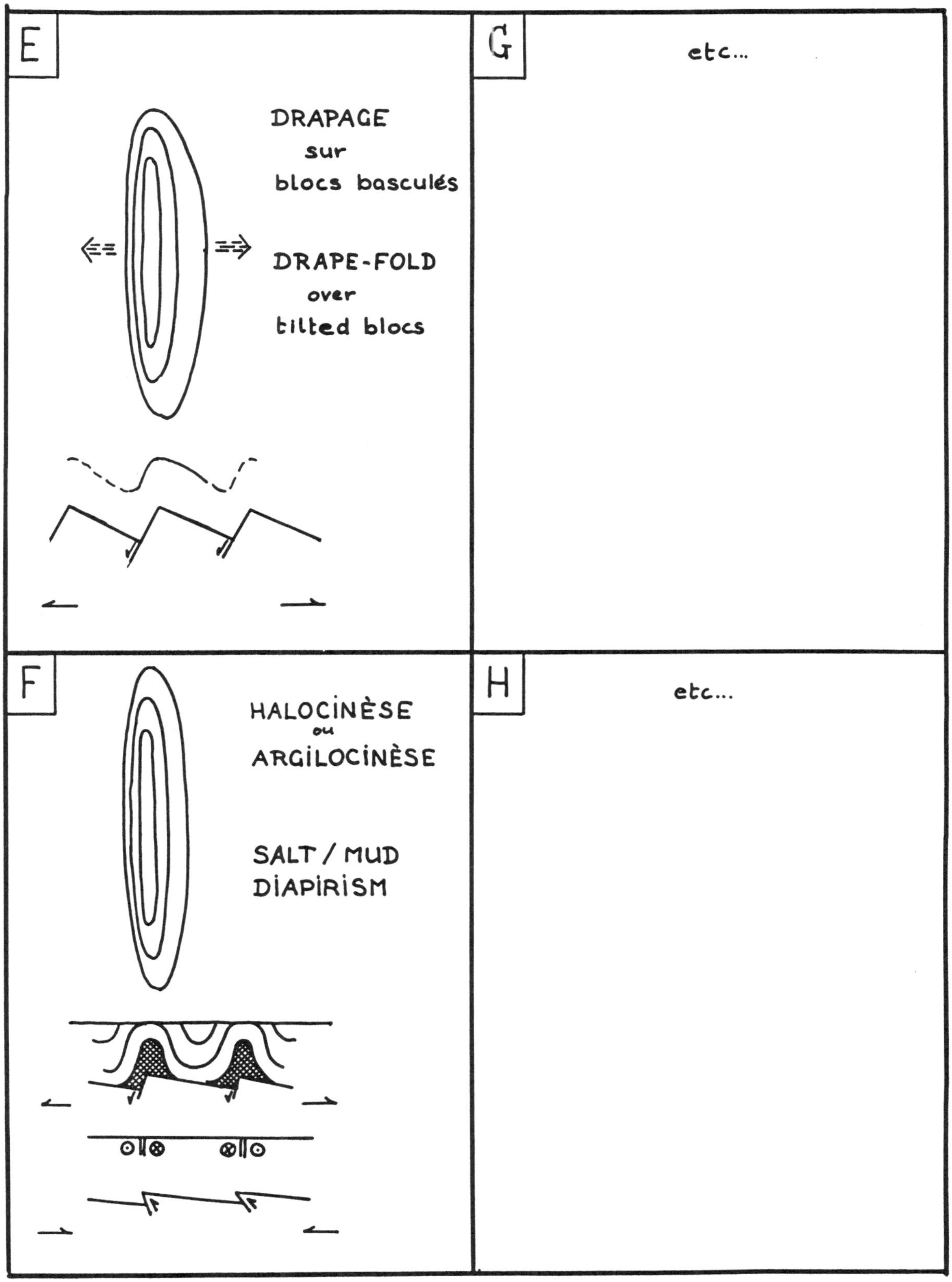

Fig. VIII-3/C Solutions

Fig. VIII-3/C Solutions

EXERCISES - EXERCICES

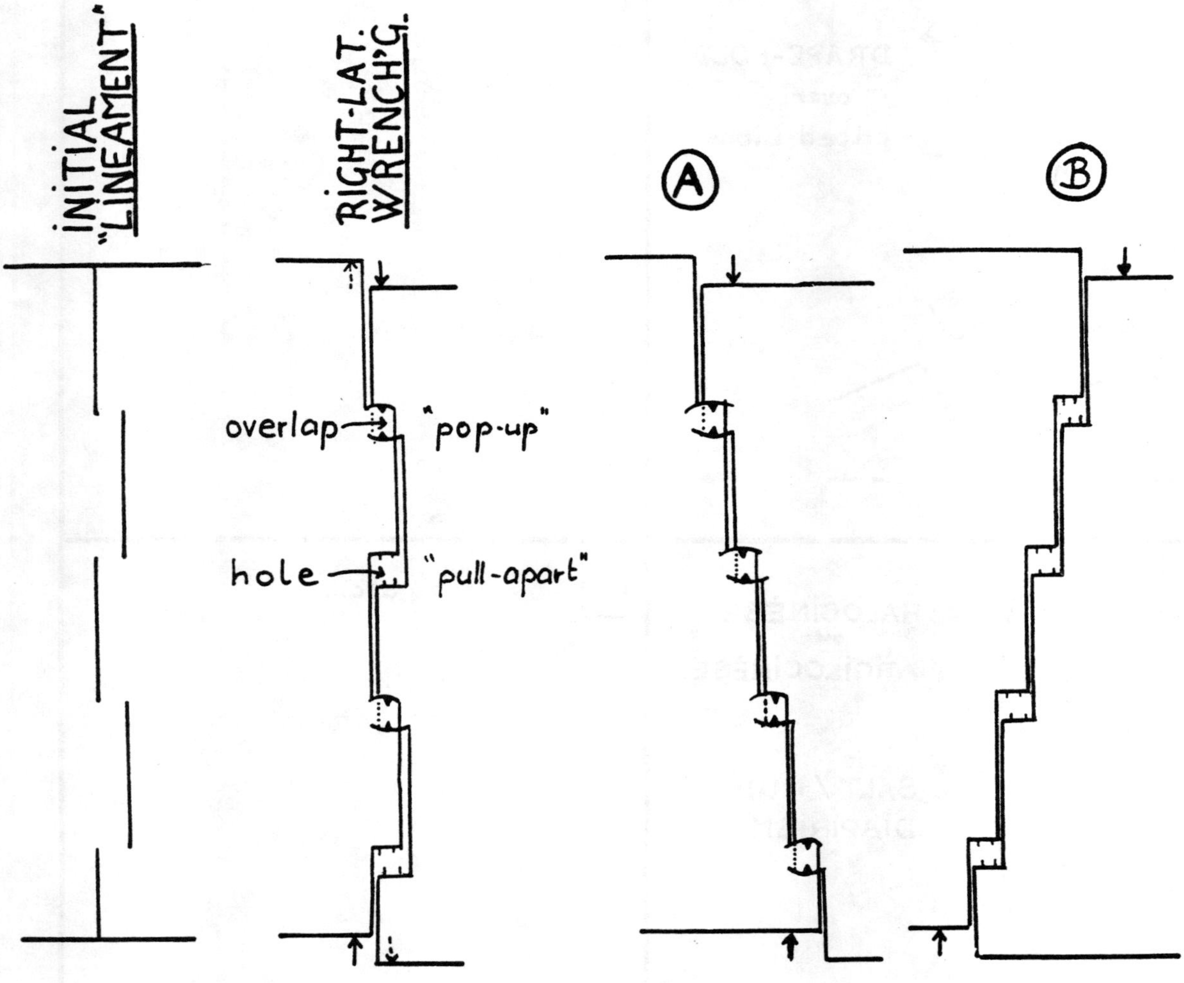

Fig. VIII-4/A "Where is the MASS-BALANCE gone ?" *Fig. VIII-4/A*

"Qu'est donc advenu au BILAN des VOLUMES ?"

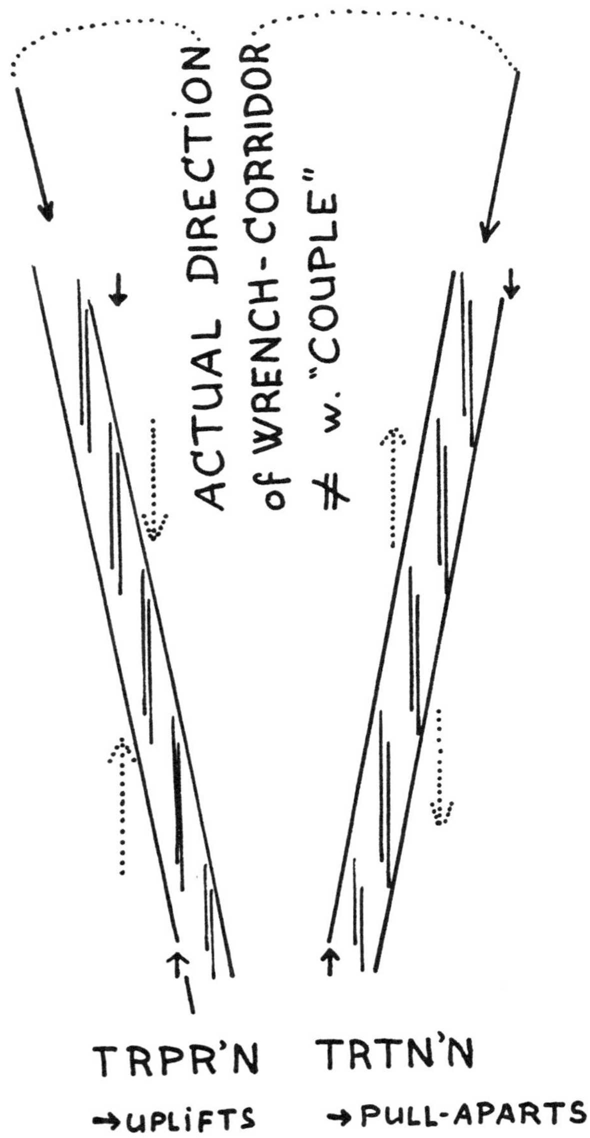

Fig. VIII-4/B Solution

Fig. VIII-4/B Solution

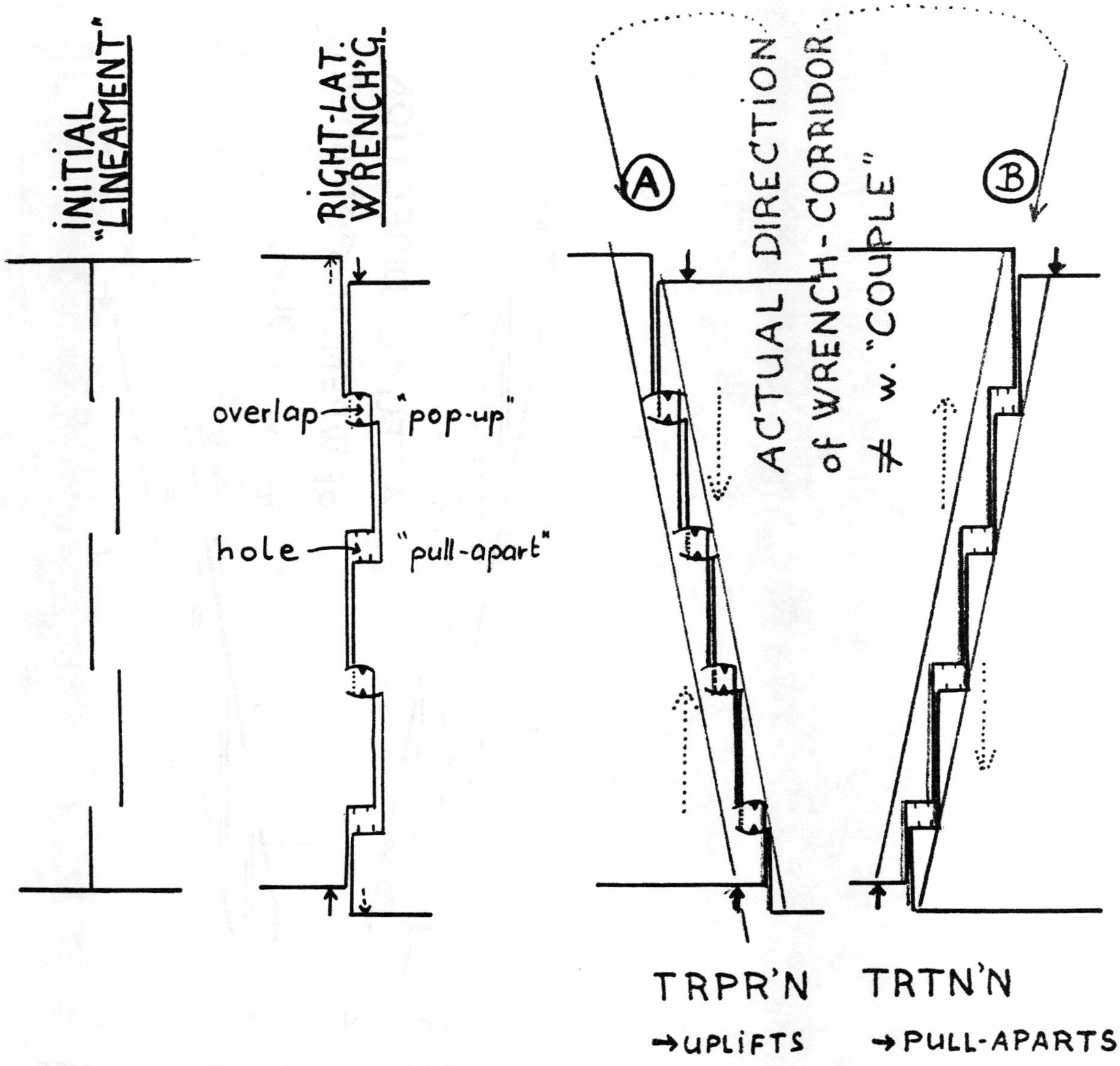

Fig. VIII-4/C Solution

Fig. VIII-4/C Solution

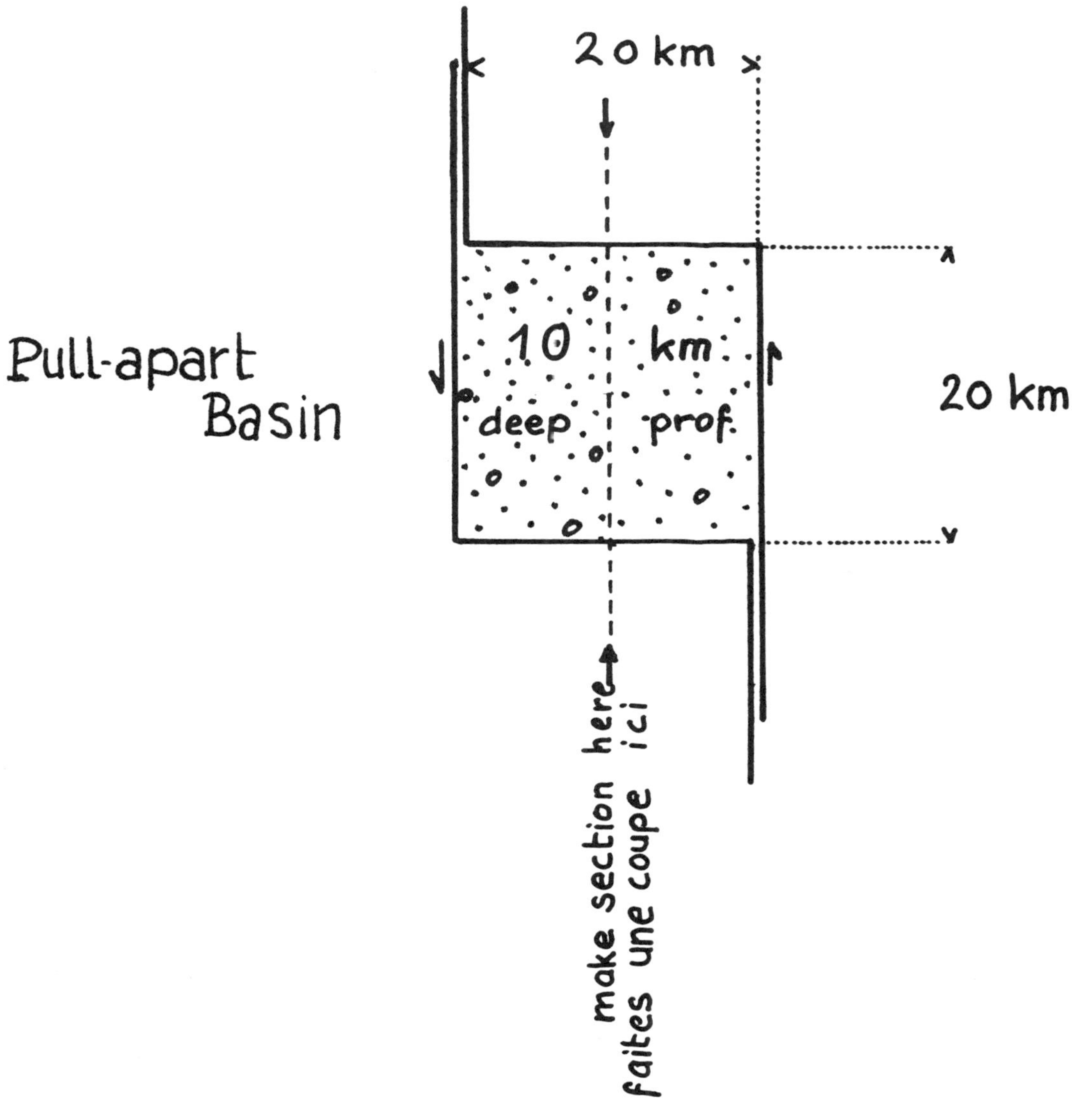

Fig. VIII-5/A THINNING of the crust through STRETCHING *Fig. VIII-5/A*

AMINCISSEMENT de la croûte par ETIREMENT

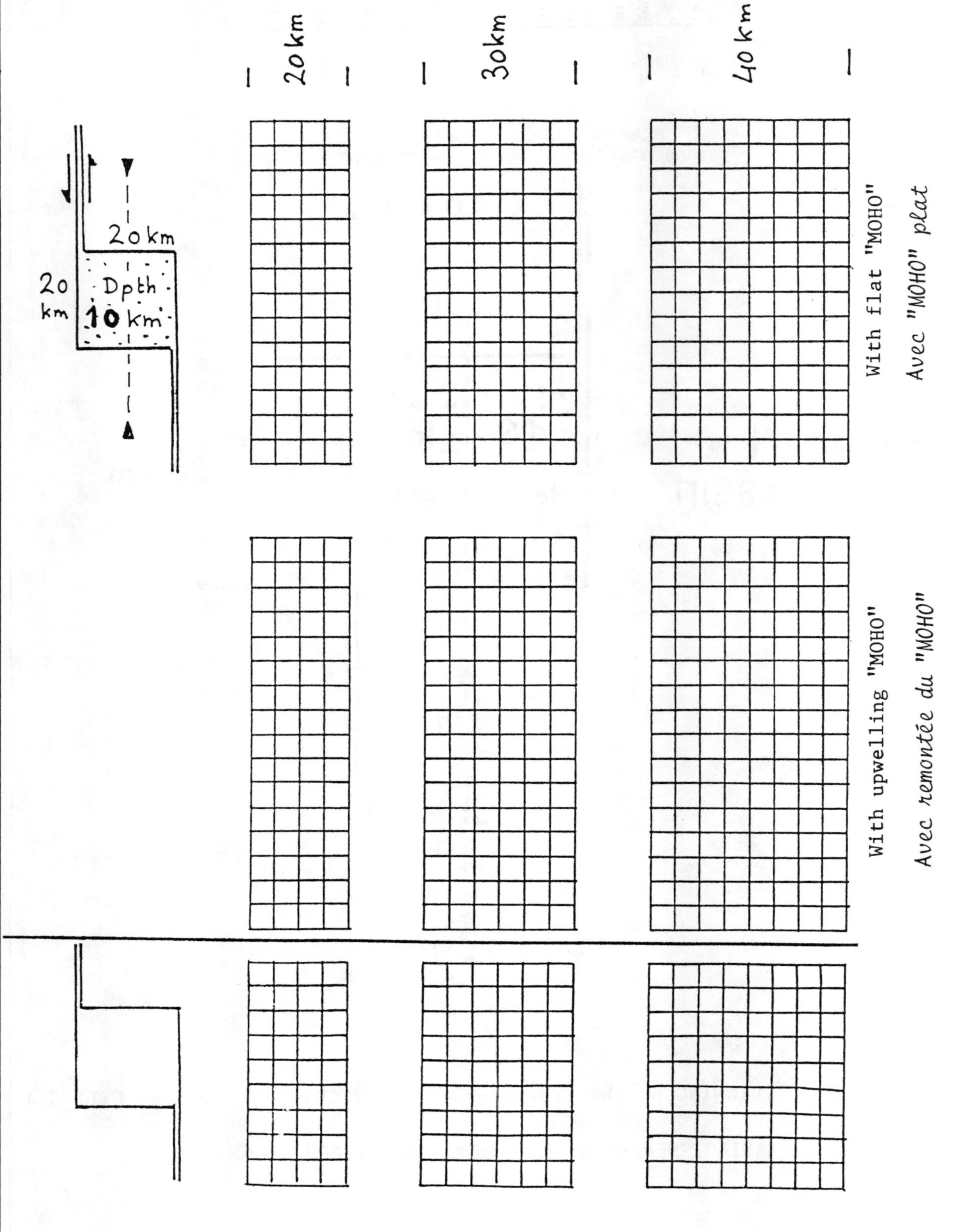

Fig. VIII-5/B Working sheet

Fig. VIII-5/B Feuille de brouillon

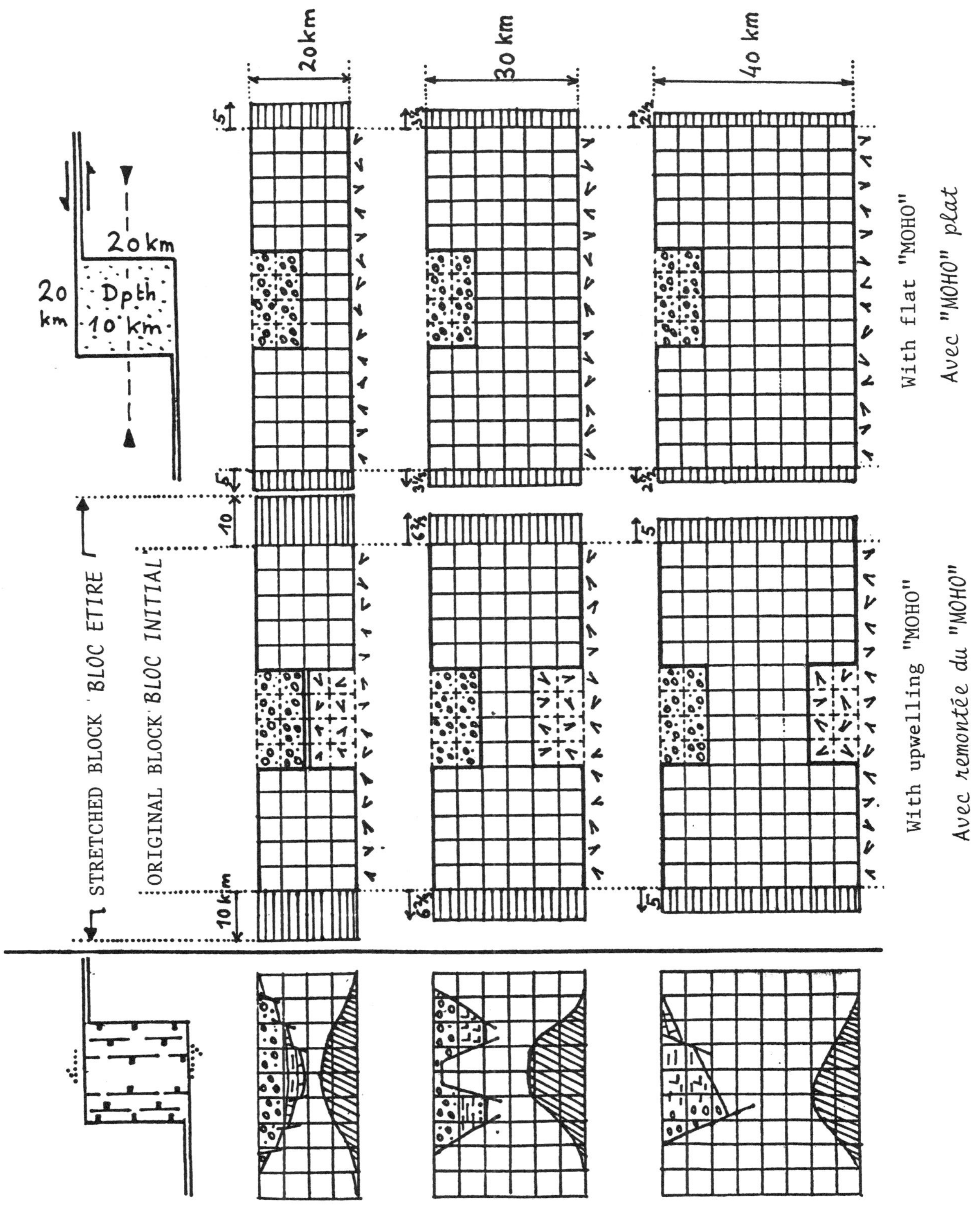

Fig. VIII-5/C Solutions

Fig. VIII-5/C Solutions

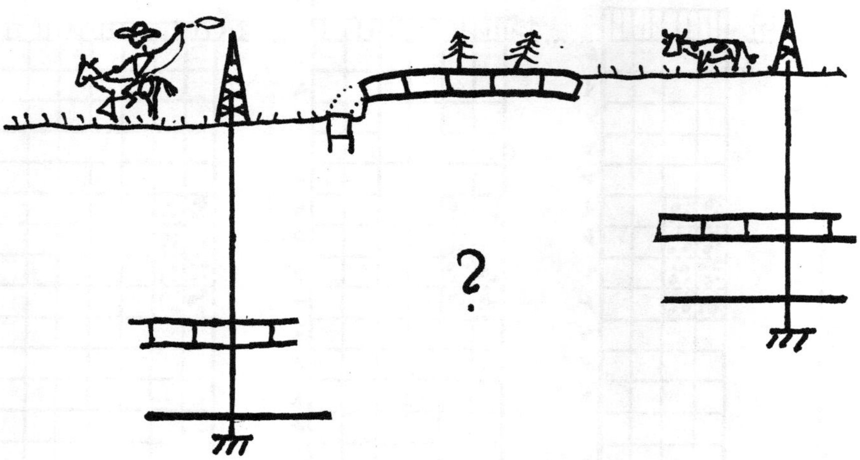

(a) drape-fold

(a) pli de drapage

Fig. VIII-6/A Flattening out of folds, or "HOW MUCH IS TOO MUCH ?"
Fig. VIII-6/A "Jusqu'où peut-on (ne doit-on pas) ALLER TROP LOIN ?"

(b) drag-folds

(b) pli d'entraînement

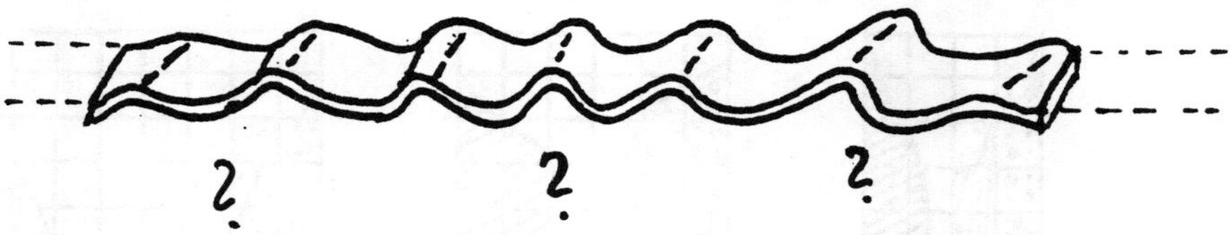

Fig. VIII-6/B Solutions

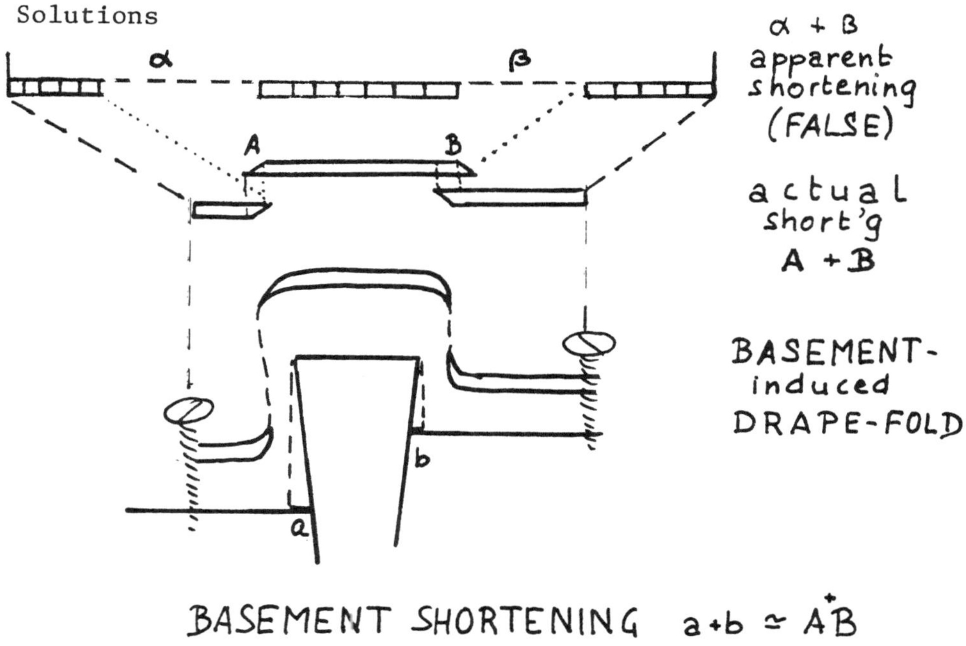

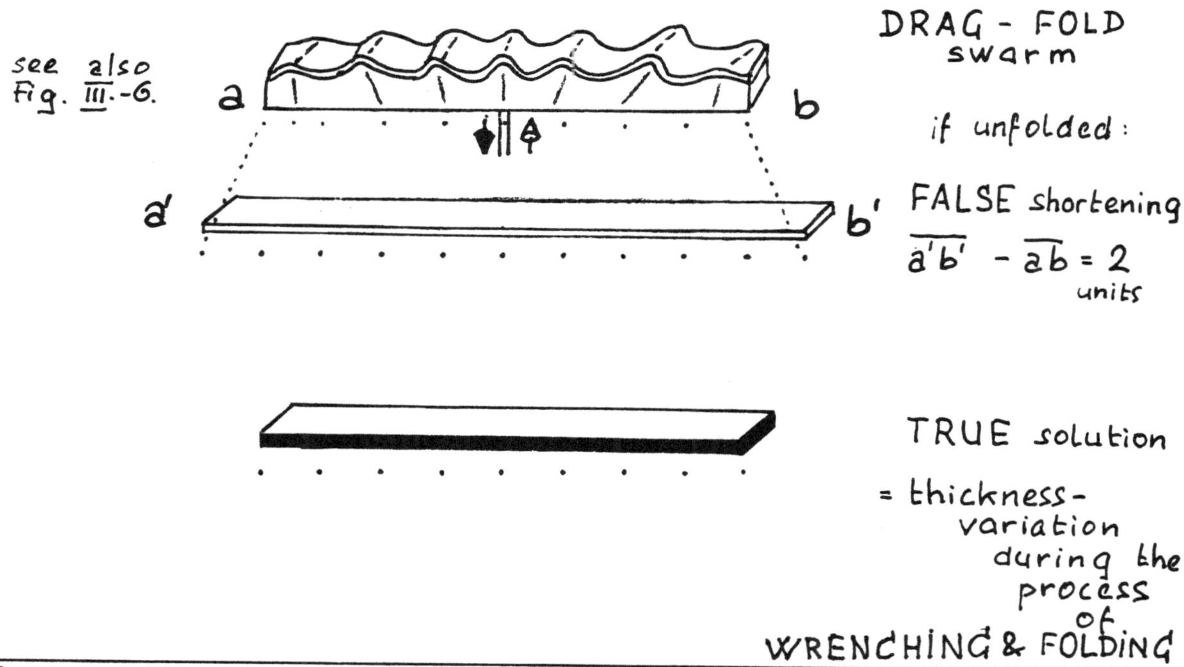

Message : Why not to un-fold drape and/or drag-folds ,

NO PALINSPASTICS should be applied in WRENCH-DOMINATED areas.

Message : *Pourquoi on n'a pas le droit de "déplier" les plis de drapage ou plis d'entraînement ?*

Il NE faut PAS FAIRE des restitutions PALINSPASTIQUES dans des zones à déformations DECROCHANTES.

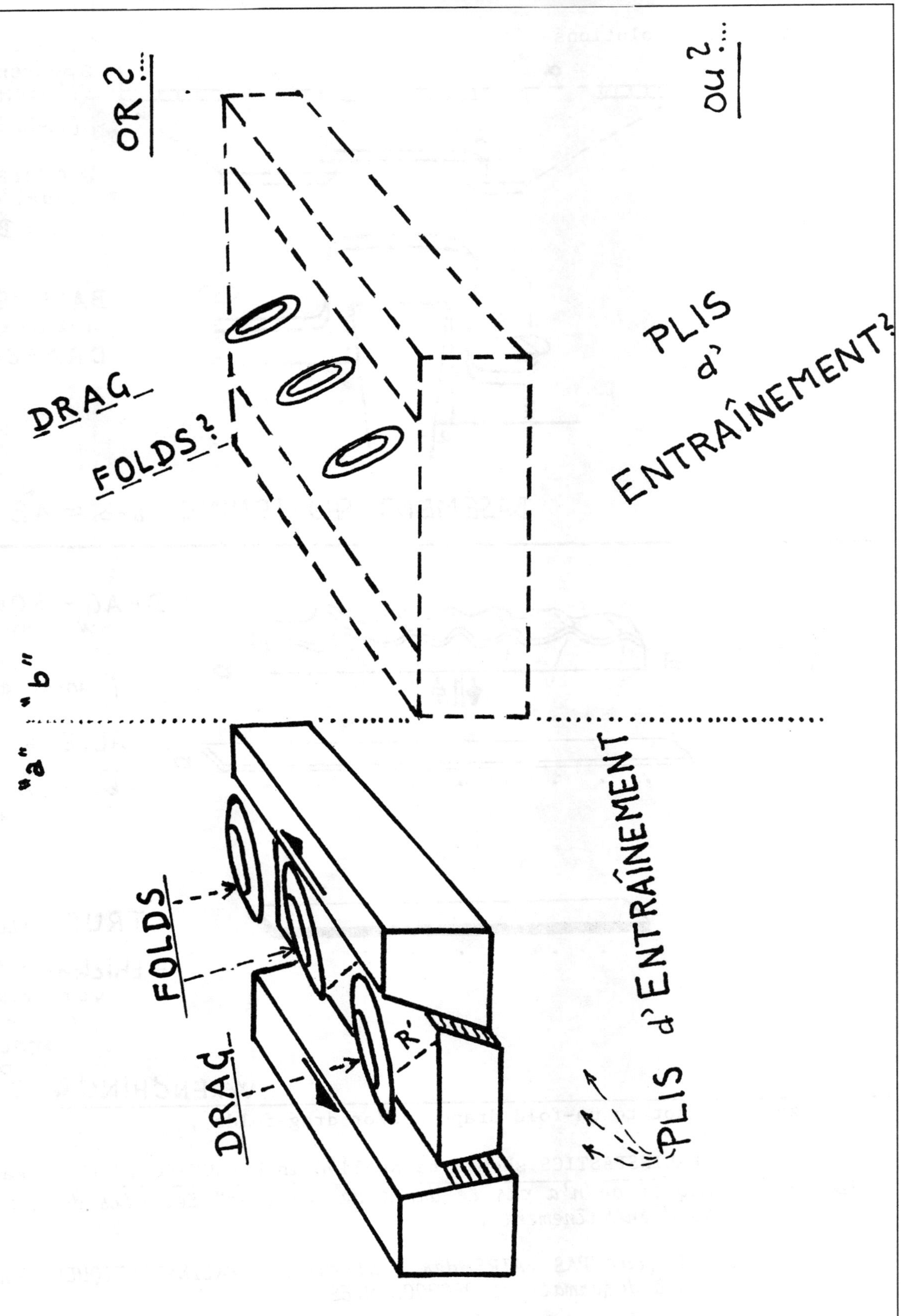

Fig. VIII-7/A Drape-fold or drag-fold

Fig. VIII-7/A Drapage OU entraînement ?

PLIS de DRAPAGE sur BLOCS de SUBSTRATUM basculés !!!

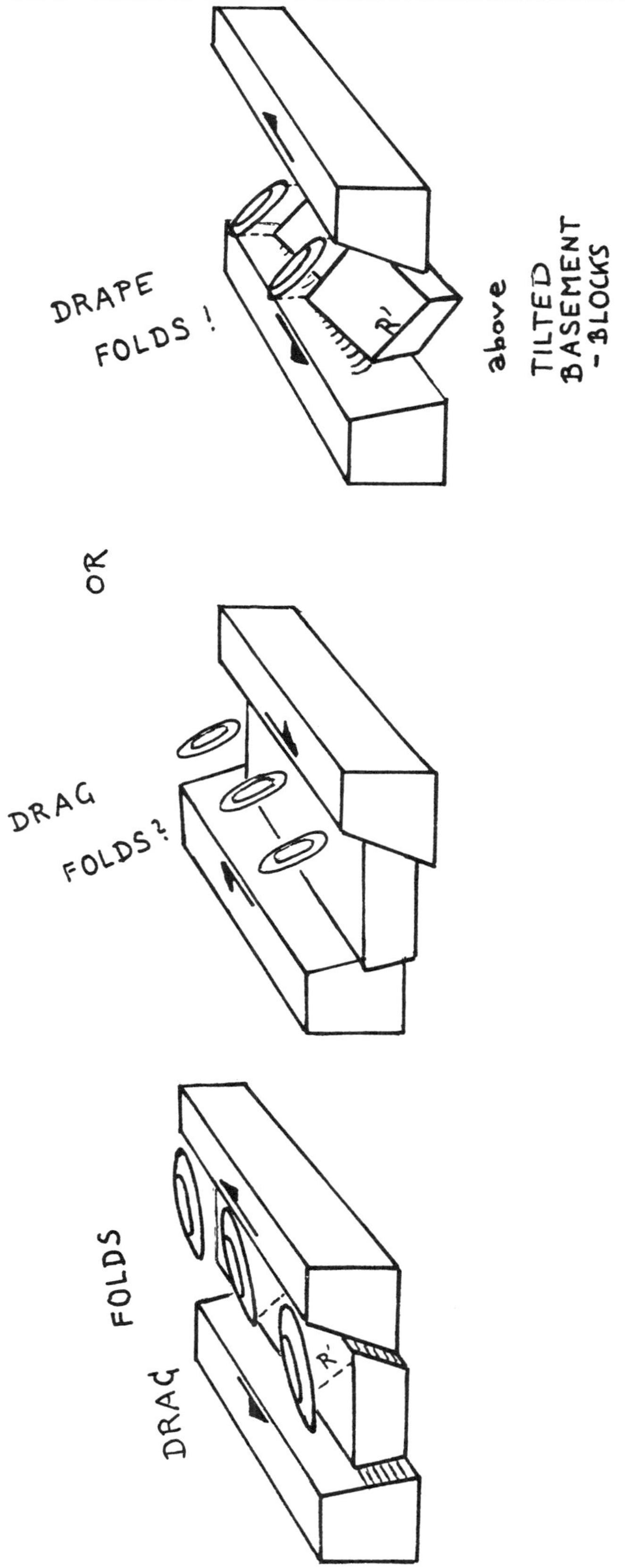

Fig. VIII-7/B Solutions

Fig. VIII-7/B Solutions

EXERCISES - EXERCICES

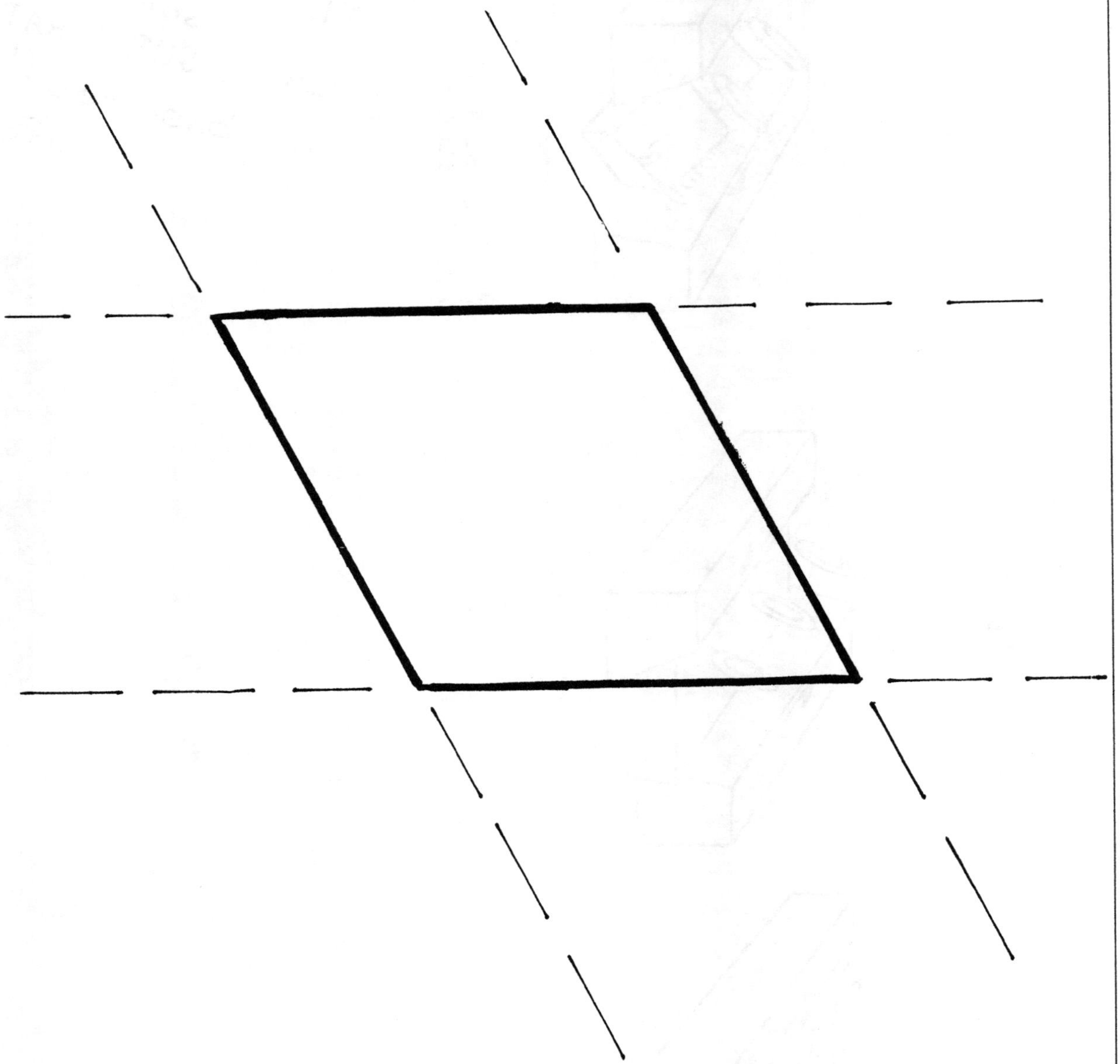

Fig. VIII-8/A LOZENGE - BASINS

Fig. VIII-8/A BASSINS LOSANGIQUES

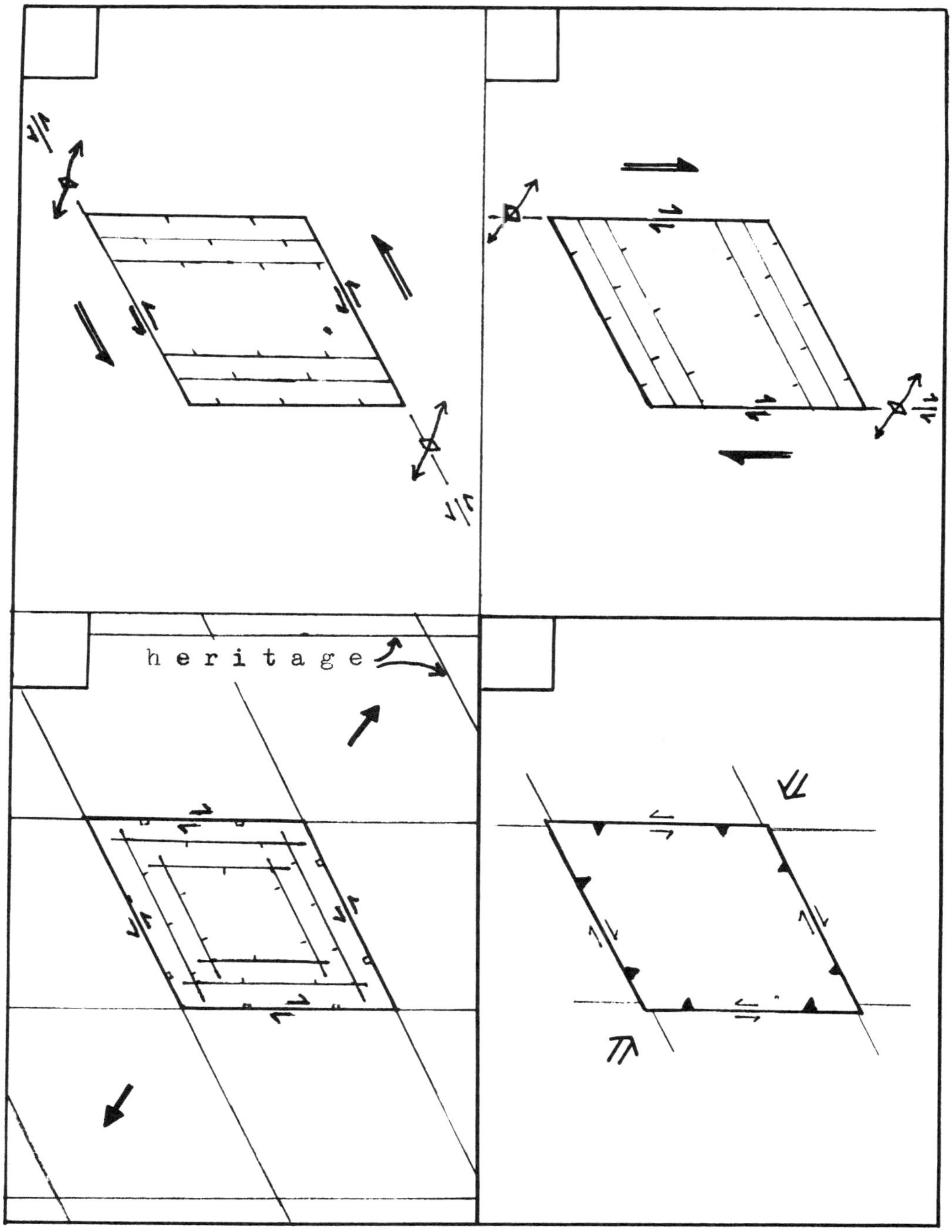

Fig. VIII-8/B Solutions

Fig. VIII-8/B Solutions

EXERCISE

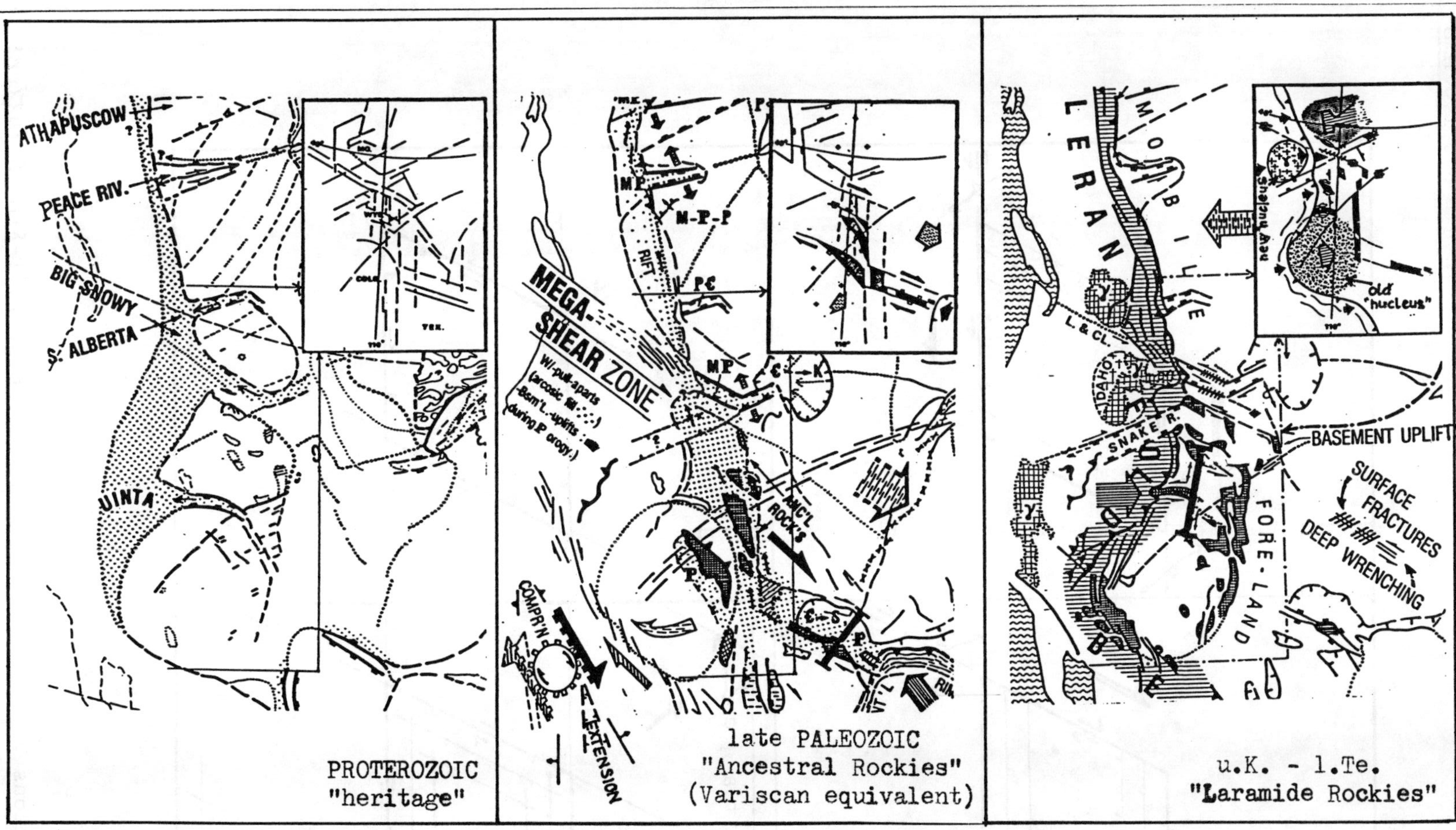

Fig. VIII-9 "SCISSOR - and - SCOTCH-TAPE exercice"

"ANCESTRAL" vs. "LARAMIDE" Rockies

Fig. VIII-9 "Exercice avec des CISEAUX et du RUBAN ADHESIF"

Des Rocheuses "ANCESTRALES" aux Rocheuses "LARAMIENNES"

FIGURES

IX. GLOBAL INTERPRETATION AND DRIVING MECHANISMS

INTERPRETATION GLOBALE ET MECANISMES PROFONDS

IX-B) FIGURES

IX-A) FIGURES

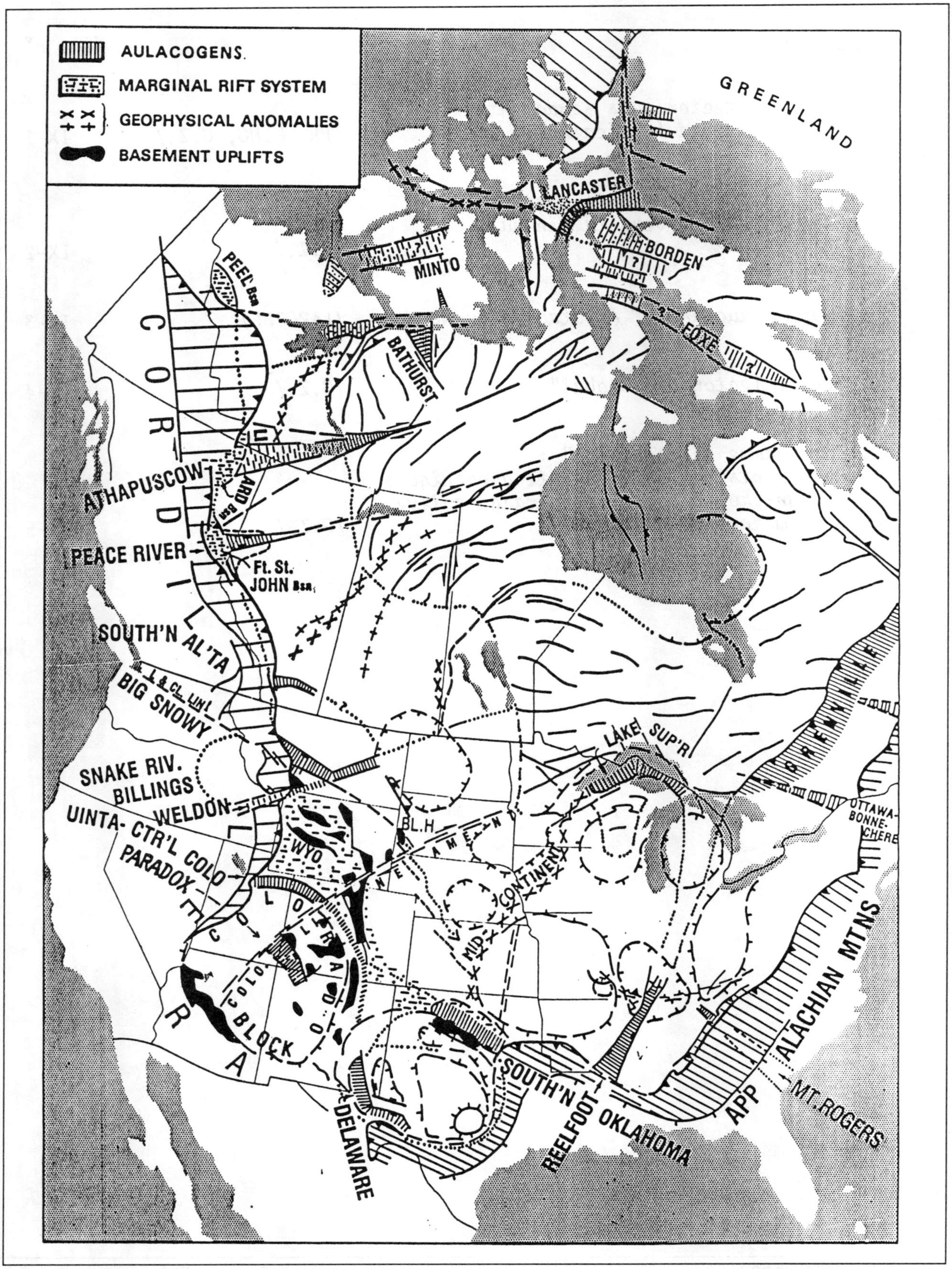

Fig. IX-1 - Basic structural elements - North America

- *Eléments tectoniques fondamentaux - Continent Nord-Américain* (Ph. KING, G.Z.)

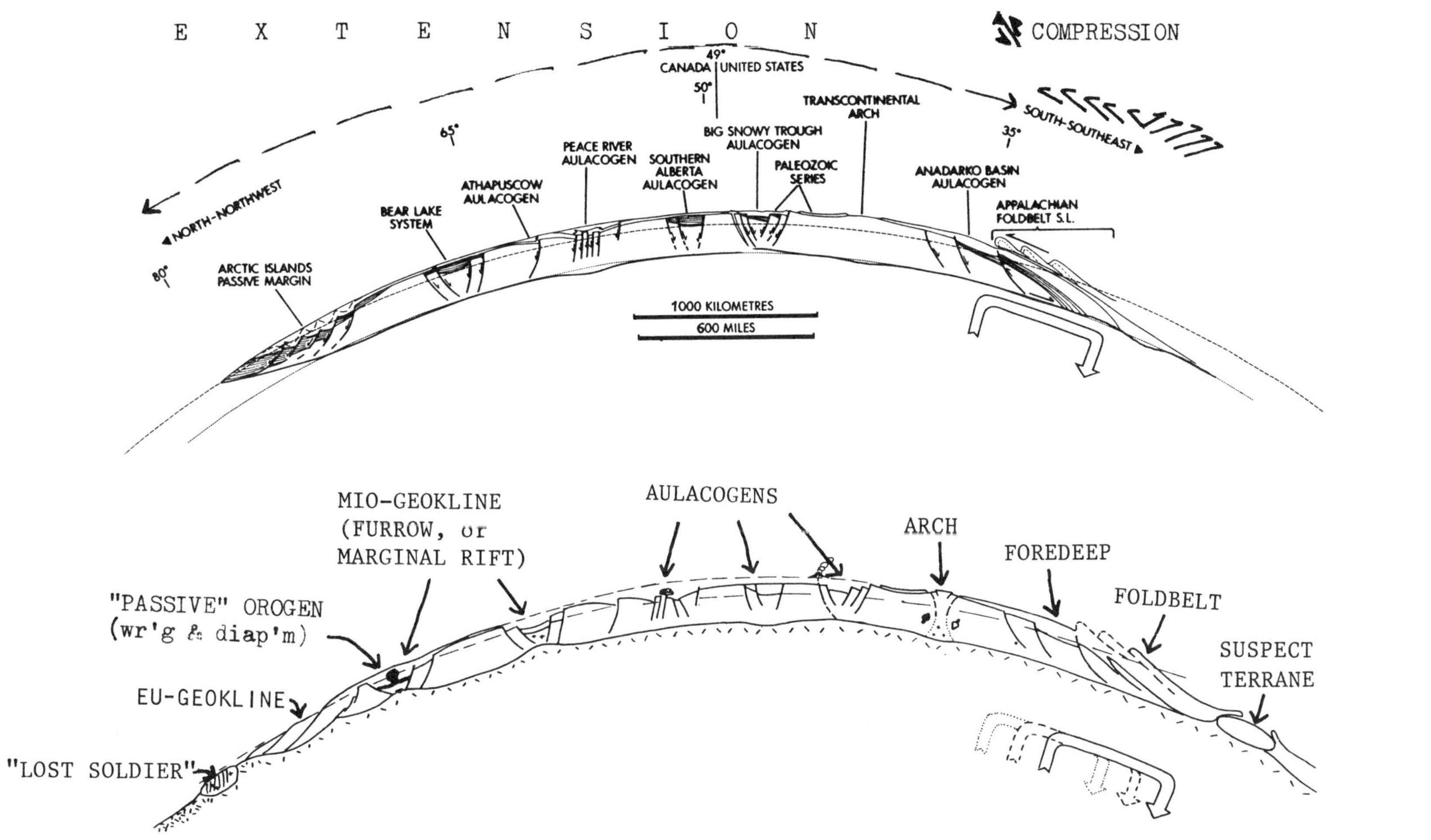

Fig. IX-2 - N-S cross-section of the North American Continent, during the late Paleozoic
(G.Z.1982)
- *Coupe N-S à travers le Continent Nord-Américain vers la fin du Paléozoïque*
(G.Z. 1982)

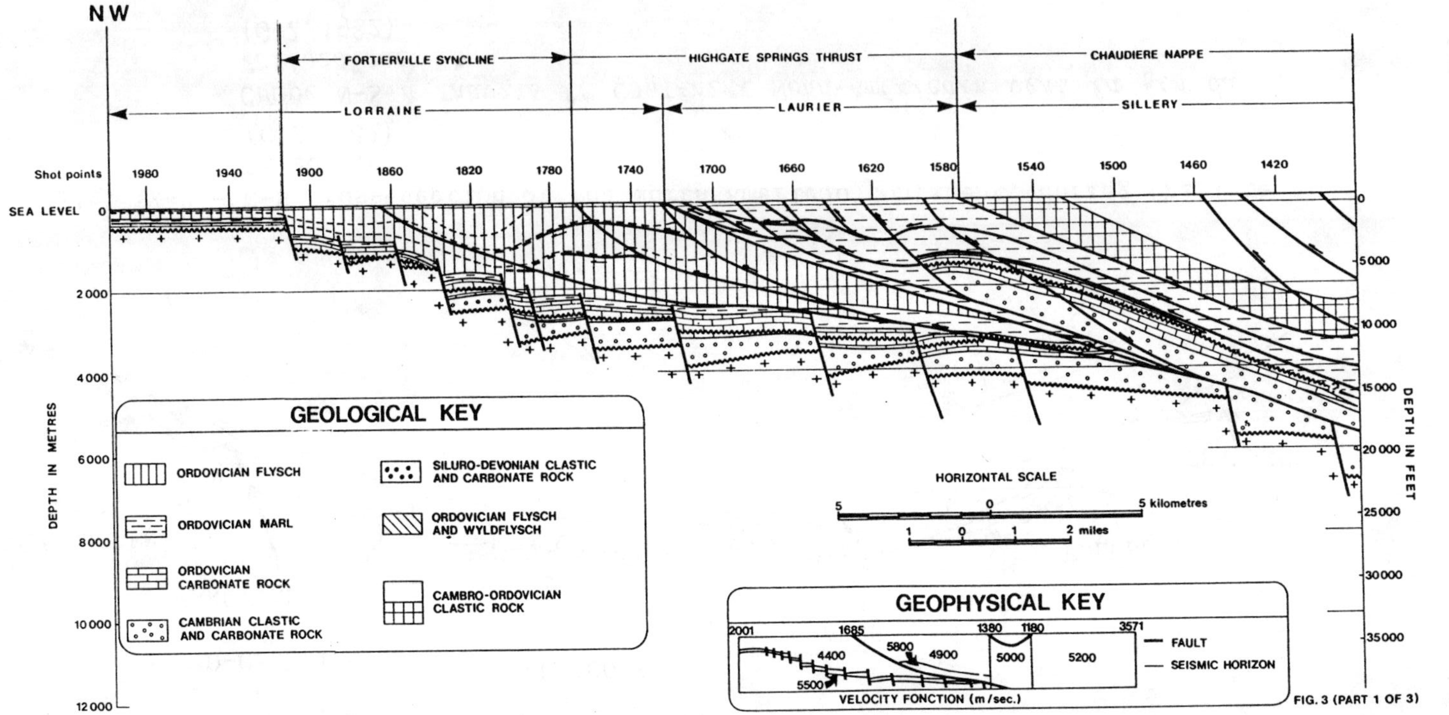

FIG. 3 (PART 1 OF 3)

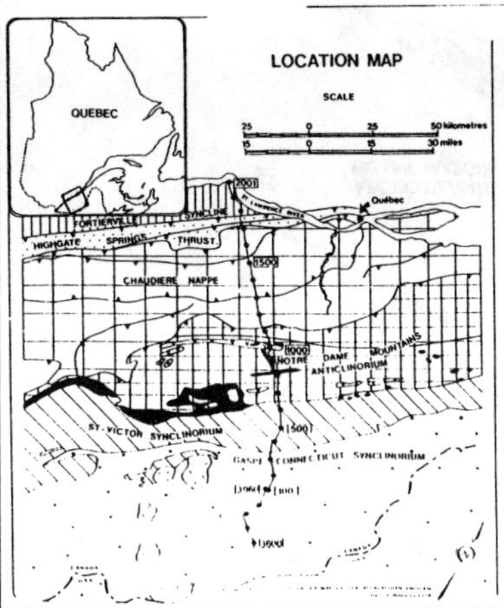

Note normal-faulting directly underneath of the Overthrust system.

Observez les failles normales directement en-dessous des Chevauchements.

Fig. IX-3 - North Appalachian cross-section
(LAROCHE)

- *Coupe structurale, Appalaches du Nord*
(LAROCHE)

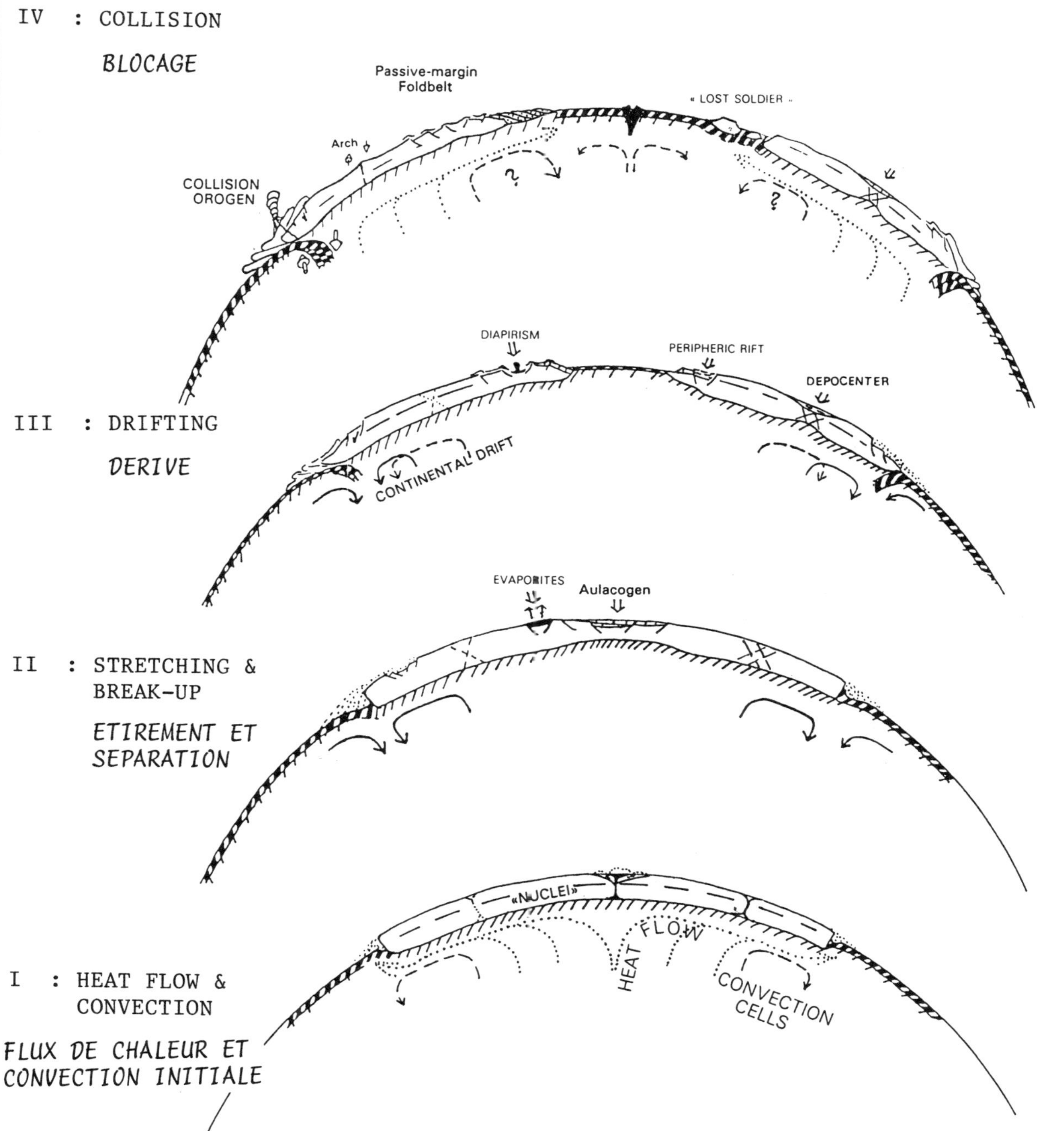

Fig. IX-4

"GLOBAL TECTONIC" cycle

Cycle de "TECTONIQUE GLOBALE"

Fig. IX-4

Fig. IX-5 Diagramme de Collision Continent-Continent.
(A) Evénements anté-orogéniques (B) Evénements syn-orogéniques

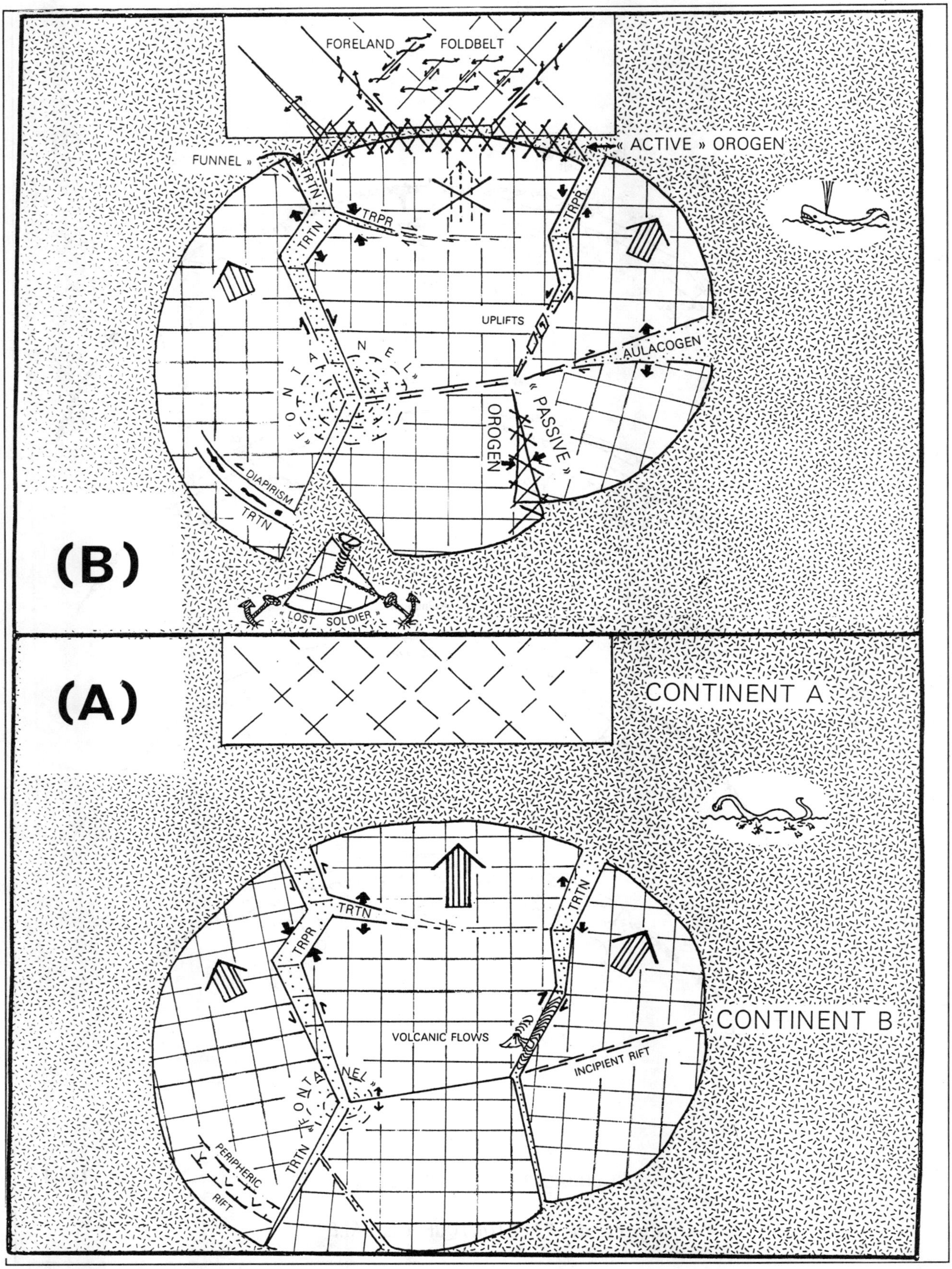

A.A.P.G. - NICE - 1988

Short course n° 2 G.Z.

Fig. IX-5 Continent-Continent Collision-diagram
(A) Pre-orogenic events (B) Syn-orogenic events